FIRST EDITION

FIRST PRINTING—1973

International Standard Book Number: 0-672-21022-3
Library of Congress Catalog Card Number: 73-85140

Questions and Answers About Noise in Electronics

by

Courtney Hall

HOWARD W. SAMS & CO., INC.
THE BOBBS-MERRILL CO., INC.
INDIANAPOLIS · KANSAS CITY · NEW YORK

Preface

Literature on electronic noise ranges from volumes describing its complex mathematical nature to brief discussions in design and reference texts. Other works delve deeply into the theoretical causes and analysis of noise in various electronic devices. The average engineer, technician, or experimenter who needs to quickly acquaint himself with the practical aspects of noise and its measurement may spend considerable time in the study of reference material, yet still not be aware of some of the important characteristics and practical measurement techniques.

This work is an attempt to provide the reader with a basic understanding of noise characteristics and the measurement peculiarities associated with it. Emphasis is placed on clear explanations of practical situations. Where mathematics become somewhat tedious, examples and tables of solutions are given so the reader may quickly understand the principles and see the effects without performing the operations. It is hoped that this book will not only introduce the reader to the subject of noise, but that it will assist those embarking on an in-depth study of its manifestations.

I wish to thank Mr. Jim Fisk for his helpful suggestions regarding selection of some of the subject matter.

COURTNEY HALL

Contents

PART 1

PART 2

PART 3

PART 4

PART 5

PART 6

PART 7

PART 8

1

Introduction

1–1. *What is noise?*

Small random voltages and currents exist in electronic components and circuits. These fluctuations are referred to as noise because they generally serve no purpose and can interfere with the measurement or detection of useful signals. Such noise sources are inherent in most electronic components and cannot be eliminated. The two major types of circuit noise are thermal noise and shot noise.

Sources of electronic noise also exist in nature, outside of electronic circuits. Atmospheric noise and galactic noise may be heard on a receiver as an increase in noise output of the receiver, when an antenna is connected to it.

The above types of noise are natural, as opposed to man-made. Most man-made noise does not have the random characteristics of natural noise and is often called interference. Man-made noise may be reduced or eliminated by filtering or shielding at the noise source.

Hum caused by power-supply ripple may also be referred to as noise in high-fidelity music systems, but it bears no similarity whatever to natural noise. It is in the class of man-made interference, and it can usually be dealt with effectively by proper circuit design and shielding.

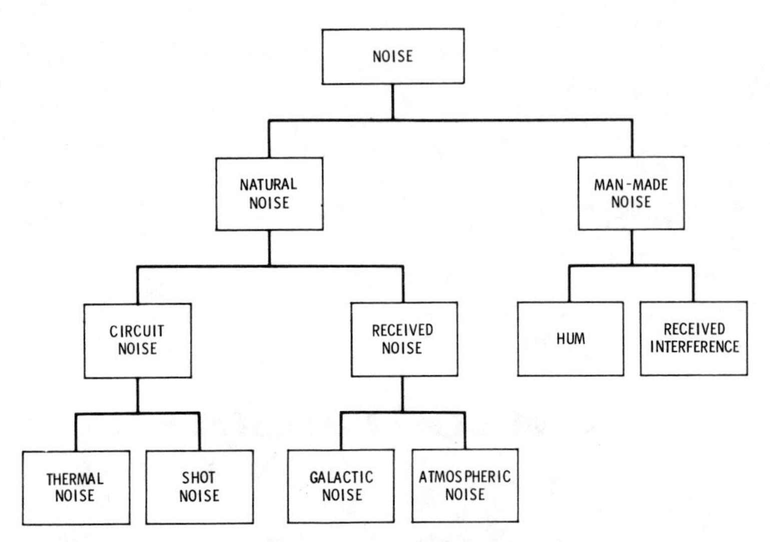

Fig. 1-1. Major types of electronic noise.

Fig. 1-1 is a chart showing how the major types of noise may be categorized.

1–2. *Why is noise important?*

Noise has little importance in an electronic system when signals are much larger or stronger than the noise. If, however, the signal is very small or weak, the noise may be sufficient to render the signal undetectable. Thus noise becomes very important when dealing with very low-level signals. Noise is the limiting factor in determining the maximum sensitivity of an electronic system.

Many types of electronic equipment must be designed to detect or measure extremely small signals. Radio and television receivers, sensitive electronic test equipment, and electronic medical instruments are just a few of the applications in which noise can be a vital parameter.

1–3. *How can the effects of noise be easily demonstrated?*

Tune a television set to one of the higher numbered uhf channels, well removed from any channels used by local stations. Turn up the volume and listen to the hissing or rushing sound. This sound is caused by circuit noise in the electronic components of the set. It is not unlike the sound of a large waterfall. Although the noise is quite small in amplitude where it originates, the noise associated with the input circuits of the set is amplified by the entire gain of the receiver and is easily heard. Proof that the noise originates inside the television set may be demonstrated by disconnecting the antenna and observing that the noise level does not change.

The visual presentation on the television picture tube is also produced by electrical noise. It appears as a random occurrence of constantly changing black, white, and gray spots having no regular pattern. In fact, it resembles a dense swarm of small insects flying about with no coherent direction. This is characteristic of the random nature of noise voltages and currents.

Another important characteristic of circuit noise is that its amplitude remains constant regardless of frequency. This may be demonstrated by varying the channel selector of the television set and observing that the sound and visual display do not change in intensity.

Television viewed in fringe areas may appear "snowy." This is because the received television signal is not large enough, compared to the noise, to produce a clear, sharp picture.

1–4. *What is white noise?*

The term "white noise" is often used to indicate that the noise under discussion has a constant value of average amplitude versus frequency. This stems from the idea that white light contains all colors of the visible spectrum, so white noise must contain all frequencies of the electronic spectrum. White noise is also said to be "flat" with frequency. Thermal noise and shot noise do not change in average amplitude as frequency is varied, so they are examples of white noise.

1–5. *What is pink noise?*

Pink noise is white noise that has been filtered so that the lower frequency components of the noise are larger than the higher frequency components. The word "pink" derives from the fact that the color pink is a mixture of white and red colors; since red is at the low-frequency end of the visible spectrum, then pink noise must be richer in the low-frequency noise than it is in high-frequency noise. Fig. 1-2 shows a typical graph of pink noise amplitude versus frequency.

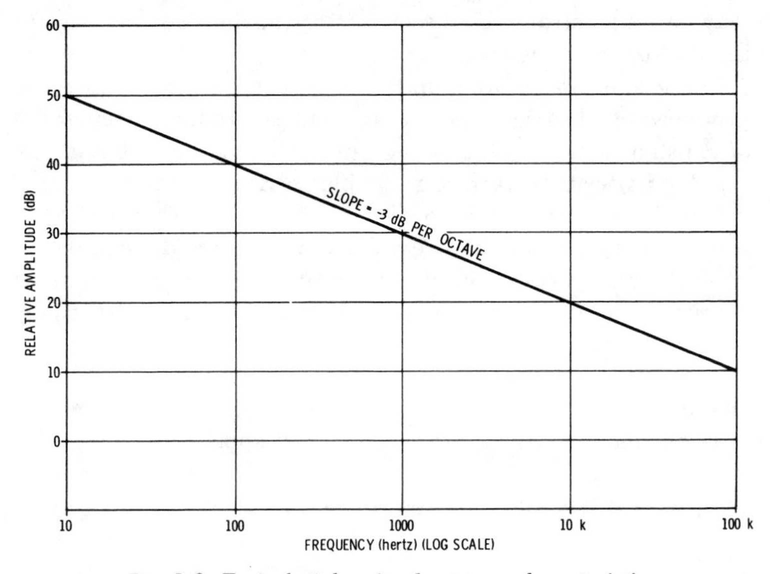

Fig. 1-2. Typical pink noise frequency characteristic.

Pink noise is used in some electronic sound synthesizers. It can also be used to provide a constant-amplitude, narrow-bandwidth noise voltage when fed through a variable-frequency bandpass filter with a bandwidth that is always a fixed percentage of the filter center frequency. Such a filter has a decreasing bandwidth as frequency is lowered. Since noise amplitude depends on bandwidth, the tendency for the noise output to decrease at low frequencies, due to decreasing bandwidth, is compensated for by the increasing amplitude of pink noise.

1–6. *What is man-made noise?*

Man-made noise is any electrical interference that originates in equipment or machinery such as automobile ignition systems, electric motors, fluorescent lamps, neon signs, x-ray and diathermy equipment, and electric welding equipment. Transient electrical disturbances caused by loads being switched on and off of the power line will also generate interference, as will poor connections in high-power circuits. In general, man-made noise is likely to be generated whenever an electrical spark or discharge occurs. Although the frequency of occurrence of these disturbances may be quite low, their harmonic content is usually very high, resulting in a broad spectrum of interference. The average amplitude of man-made noise tends to decrease as frequency increases. As would be expected, its amplitude is considerably less in rural areas than in large cities.

Noise limiters in communications receivers can provide a degree of relief from man-made noise. They are most effective when the interference is in the form of narrow pulses, with relatively long periods between pulses. The best cure for man-made noise is to prevent its radiation from the point of origination by shielding and filtering, but this is impractical in large cities where the sources of man-made noise are numerous. The most practical solution is to move the receiver to a quiet location, away from noise sources.

1–7. *What is atmospheric noise?*

Atmospheric noise is caused by lightning in thunderstorms. The peak powers radiated by lightning discharges are tremendous, and their effects may propagate over very great distances. The level of atmospheric noise may widely vary with season, time of day, and geographical location, but it can be expected to be higher in summer months when nearby thunderstorm activity is greatest.

Fig. 1-3 shows a very approximate relation between atmospheric noise and frequency. Received noise in microvolts is based on a 50-ohm nondirectional antenna and a receiver bandwidth of 5 kHz. Smaller bandwidths will result in lower noise levels. Although the actual values of noise voltage versus frequency are highly variable, the general shape of the curve is of interest. It shows that typical circuit noise in receivers is negligible, compared to atmospheric noise at frequencies below 30 MHz or so. Thus, there is little, if any, advantage in extremely low-noise receivers for the lower frequencies.

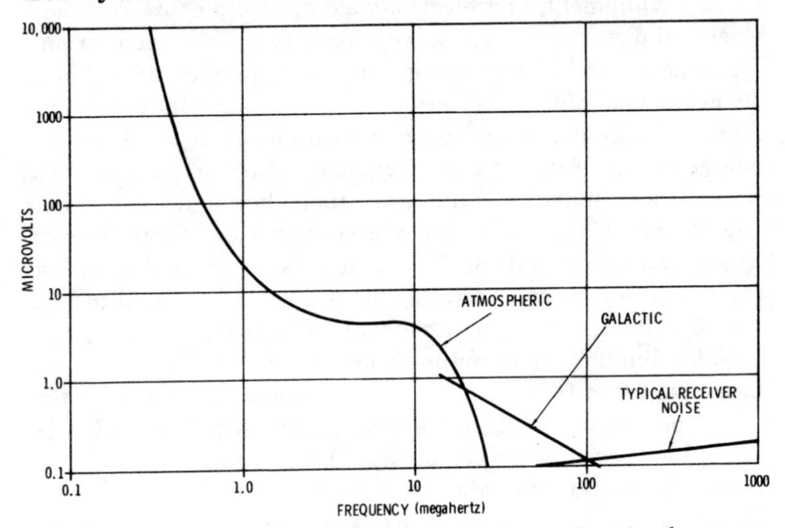

Fig. 1-3. Received noise vs frequency for a 50-ohm nondirectional antenna and a 5-kHz receiver bandwidth.

Atmospheric noise may be heard on sensitive communications receivers tuned to unused frequencies below about 25 MHz. The increase in noise heard when the antenna is connected to the receiver is due to atmospheric noise picked up by the antenna. As indicated by Fig. 1-3, the amplitude of atmospheric noise increases as the receiver frequency is decreased. A failure to detect atmospheric noise by this method may indicate that the receiver is lacking in sensitivity.

1–8. *What is galactic noise?*

Galactic noise originates outside the earth's atmosphere from such sources as the sun, our galaxy, the Milky Way, and other galaxies. The ionosphere prevents galactic noise frequencies below about 15 MHz from reaching the earth, and typical receiver noise limits its reception to frequencies less than a few hundred MHz. Fig. 1-3 shows approximate galatic noise levels for a 50-ohm nondirectional antenna and a receiver bandwidth of 5 kHz.

Galactic noise may be detected with a sensitive television receiver, if local stations do not use channels 2 or 3. By setting the channel selector to channel 2 (54 to 60 MHz), galactic noise may be heard as an increase in noise when the antenna is connected to the set.

Radio astronomers detect and measure galactic noise in their study of outer space. Their radio telescopes, employing large, high-gain antennas and very low-noise receivers, can detect galactic noise at frequencies higher than those indicated in Fig. 1-3.

2

Thermal Noise

2–1. *What is thermal noise?*

Thermal noise, also called Johnson noise or resistance noise, occurs naturally in all electrical resistors and in all electronic components which contain resistance. It is not necessary for a current to flow in the resistor for thermal noise to exist. It appears across the terminals of an open-circuited resistor; thus, it may be treated as a noise voltage generator. Fig. 2-1 shows the classical equivalent thermal noise circuit for a resistor. The thermal noise voltage generator is in series with a noiseless resistor, having the same ohmic value as the actual resistor. Thermal noise has no dc component; it is strictly an ac voltage.

Thermal noise is caused by the random motion of free electrons in a resistance. As the temperature of the resistor is increased, the agitation of these electrons becomes more vigorous, raising the noise voltage. Hence, the name thermal noise is given. If the temperature of the resistor could be lowered to absolute zero (−273 °C), then the value of its thermal noise voltage would be zero.

The amplitude of thermal noise depends on the value of the resistance; the larger the resistance is in ohms, the larger the thermal noise voltage. Although the value of thermal noise voltage is independent of frequency, it does depend on the band-

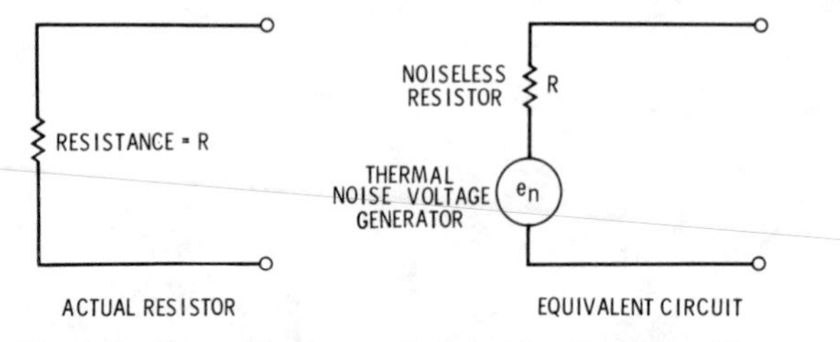

Fig. 2-1. Thermal noise equivalent circuit of a resistor.

width observed. This is illustrated in Fig. 2-2. Thermal noise voltage from the same resistor is measured after passing through four different bandpass filters. All of the filters have different center frequencies, but the bandwidths of filter 1 and filter 2 are equal to each other; so the output noise voltages, V1 and V2, are equal. Filters 3 and 4 also have equal bandwidths, and their output noise voltages, V3 and V4, are equal to each other. It is important to note, however, that filters 3 and 4 have wider bandwidths than filters 1 and 2, resulting in a higher value of noise voltage for V3 and V4 than for V1 and V2.

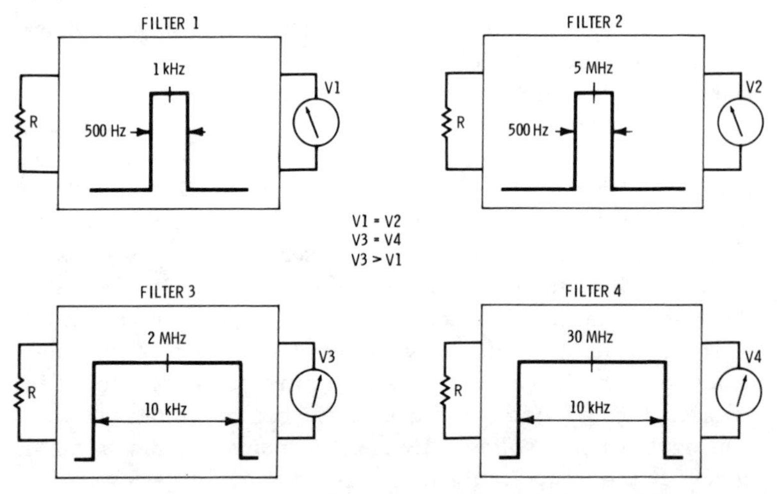

Fig. 2-2. Effect of center frequency and bandwidth on amplitude of thermal noise.

2-2. *Why is thermal noise important?*

Resistors are used extensively in electronic circuits, and their thermal noise contributes to the overall circuit noise of a system. This is especially important in the input circuits of a system, where thermal noise receives the greatest amount of amplification before reaching the output.

Signal sources, such as microphones, antennas, signal generators, and various transducers, have an internal resistance effectively in series with their output. This is called the output resistance, generator resistance, or source resistance; it is generally indicated by the notation R_g. Fig. 2-3 shows an equivalent circuit which may be used for a signal source. Two voltage generators represent the signal voltage and the thermal noise voltage associated with the source resistance. Source resistance R_g is considered noiseless in the equivalent circuit. For antennas, R_g is the sum of the radiation resistance plus any resistance due to significant losses. Microphones may have a wide range of source resistances, depending on the type, but most signal generators have a source resistance of either 50 or 600 ohms.

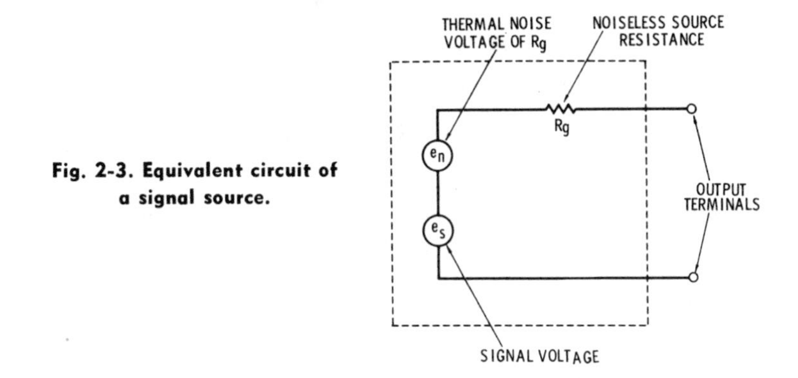

Fig. 2-3. Equivalent circuit of a signal source.

Obviously, the thermal noise voltage of the source resistance is important at small signal levels because it places a limit on the minimum usable signal voltage. This would be true even if perfect, noiseless amplifiers or receivers were available to detect the signal.

2–3. *How is the value of thermal noise voltage calculated?*

The value of thermal noise voltage in a resistance may be calculated from Eq. 2-1 as:

$$e_n = \sqrt{4KTBR} \qquad \text{(Eq. 2-1)}$$

where,

e_n is the rms value of thermal noise in volts,
K is Boltzmann's constant (1.38×10^{-23} joules/degree Kelvin),
T is absolute temperature of the resistance in degrees Kelvin,
B is noise bandwidth of the system in hertz,
R is resistance in ohms.

Some explanation regarding the temperature and bandwidth is in order. Temperature expressed in degrees Kelvin is the number of degrees above absolute zero on the Celsius (formerly centigrade) temperature scale. Absolute zero is −273 °C, so 0 °C is equal to 273 K. Temperature, in degrees Celsius, may be easily converted to degrees Kelvin by using the simple relation of Eq. 2-2.

$$K = {}^\circ C + 273 \qquad \text{(Eq. 2-2)}$$

Temperature in degrees Fahrenheit may be converted to degrees Kelvin by use of Eq. 2-3.

$$K = 273 + \frac{5}{9}({}^\circ F - 32) \qquad \text{(Eq. 2-3)}$$

Except when unusual temperature extremes are encountered, it is common practice to use a value of 290 K for temperature in Eq. 2-1. This value corresponds to temperatures of 17 °C and 62.6 °F.

Noise bandwidth B in Eq. 2-1 is an equivalent rectangular bandwidth which may differ considerably from the 3-dB bandwidth which is commonly specified in electronic systems. You will find that noise bandwidth is discussed in detail in Part 4.

In the usual case, where temperature is taken to be 290 K, Eq. 2-1 may be reduced to a simpler form by inserting the values for K and T and taking them and the factor 4 outside of the square-root sign. When this is done,

$$
\begin{aligned}
e_n &= \sqrt{4KTBR} \\
&= 1.27 \times 10^{-10} \sqrt{BR} \qquad \text{(Eq. 2-4)}
\end{aligned}
$$

and B and R are the same as in Eq. 2-1.

It is well to remember that thermal noise voltage increases not only with temperature, but with bandwidth and resistance as well. Decreasing any of these parameters will help to minimize thermal noise.

2-4. *What are typical values of thermal noise voltage?*

According to Eq. 2-1, a pure resistance will generate an infinite noise voltage if it has an unlimited or infinite bandwidth. This is theoretically true, but actual resistors have small amounts of stray capacitance and inductance—their reactances become more significant at higher frequencies, resulting in finite bandwidths. Even if it was possible to build a resistor having a resistance of 100 megohms and a noise bandwidth of 1×10^6 MHz, its thermal noise would be only 1.27 volt at a temperature of 290 K.

A more practical example is a dynamic microphone with a source resistance of 10,000 ohms. If the noise bandwidth of the microphone is 10 kHz, then the thermal noise voltage out of the microphone terminals can be found from Eq. 2-4 as follows:

$$
\begin{aligned}
e_n &= 1.27 \times 10^{-10} \sqrt{BR} \\
&= 1.27 \times 10^{-10} \sqrt{(10^4)\,(10^4)} \\
&= 1.27 \times 10^{-10} (10^4) \\
&= 1.27 \times 10^{-6} \\
&= 1.27\ \mu V
\end{aligned}
$$

This value of thermal noise voltage can be reduced by feeding the output of the microphone through a filter which has a smaller noise bandwidth than 10 kHz. An amplifier having a smaller bandwidth will also reduce the effective value of thermal noise voltage in the microphone. Noise amplitude at the output of a system is controlled by the smallest bandwidth in that system.

Another typical example is a communications receiver connected to a 50-ohm antenna. If the noise bandwidth of this receiver is 3 kHz, the thermal noise voltage from the antenna is found to be:

$$\begin{aligned} e_n &= 1.27 \times 10^{-10} \sqrt{(3 \times 10^3)\,(50)} \\ &= 1.27 \times 10^{-10} \sqrt{15 \times 10^4} \\ &= 1.27 \times 10^{-10} (3.88 \times 10^2) \\ &= 4.9 \times 10^{-8} \\ &= 0.049\ \mu V \end{aligned}$$

Still another example of interest is a television receiver connected to a 300-ohm antenna. If the noise bandwidth of the video circuit is 5 MHz, then the thermal noise voltage of the antenna would be:

$$\begin{aligned} e_n &= 1.27 \times 10^{-10} \sqrt{BR} \\ &= 1.27 \times 10^{-10} \sqrt{(5 \times 10^6)\,(300)} \\ &= 1.27 \times 10^{-10} \sqrt{15 \times 10^8} \\ &= 1.27 \times 10^{-10} (3.88 \times 10^4) \\ &= 4.9 \times 10^{-6} \\ &= 4.9\ \mu V \end{aligned}$$

Thus a television set may see a value of thermal noise voltage in its antenna which is a hundred times larger than that seen by a communications receiver in its antenna.

These examples are based on a resistance temperature of 290 K (62.6 °F). If the actual temperature differed from this value by plus or minus 63 °F (that is, if it varied from about 0 °F to 126 °F) the calculated values of thermal noise voltage could be in error by about plus or minus 6%. This is usually not too serious, however, since the value of source resistance may not be known with any greater degree of accuracy.

2–5. *What is the meaning of the term noise temperature?*

Noise temperature is a term used to indicate noise levels. It is the temperature in degrees Kelvin that a resistor would have to reach in order to have a thermal noise equal to the noise actually present. For example, suppose the thermal noise from a 50-ohm antenna is 0.05 microvolt for the bandwidth of the receiver used, but, due to atmospheric noise received, the noise from the antenna is 5.0 microvolts. The actual temperature of the antenna is assumed to be 290 K. Since the actual noise voltage from the antenna is 100 times the thermal noise voltage, the temperature the antenna would have to reach to have this much thermal noise voltage would be $(100)^2$, or 10,000 times 290 K. Thus the antenna noise temperature due to atmospheric noise is:

$$10{,}000 \times 290\ \text{K} = 2{,}900{,}000\ \text{K}$$

Noise temperature is also used to indicate the noise performance of receivers compared to the thermal noise of their source resistance. When thermal noise and actual noise voltages are known, noise temperature, T_n, may be calculated as:

$$T_n = 290\ \text{K}\left(\frac{\text{actual noise voltage}}{\text{thermal noise voltage}}\right)^2 \qquad \text{(Eq. 2-5)}$$

3

Shot Noise

3–1. *What is shot noise?*

Shot noise occurs in vacuum tubes, transistors, diodes, and other semiconductor devices due to the flow of direct current through the device. Although the average value of this direct current may be constant, there are small, random variations or fluctuations in the instantaneous value of current. These fluctuations produce an ac shot-noise current.

The amplitude of shot-noise current increases as the values of direct current and noise bandwidth become larger, but, like thermal noise, its amplitude is independent of frequency. If the output of a shot-noise generator was fed to the inputs of the four filters in Fig. 2-2, the output voltages of the filters would theoretically behave in the same manner as for thermal noise. The center frequency of the filters would have no effect on measured noise amplitude. Equal bandwidths would produce equal noise voltage outputs, and wider bandwidths would result in larger noise amplitudes. In actual practice, shot-noise generators tend to have an excess noise component at the lower audio frequencies.

3–2. *Why is shot noise important?*

The fact that shot noise occurs in devices such as vacuum tubes and transistors (which are used to amplify weak signals) means that such amplifying devices must add more noise to that already existing, because of thermal noise in the source resistance. Thus a signal can be expected to become more noisy due to shot noise when it is amplified by conventional amplifiers.

Shot noise is also important because it may be used to considerable advantage in noise generators. These instruments have predictable output noise voltages and they can simplify the measurement of noise figure by eliminating the need to know the value of noise bandwidth.

3–3. *How does shot noise differ from thermal noise?*

The major difference between shot noise and thermal noise is the manner in which they are generated. Thermal noise results from the thermal agitation of free electrons. Shot noise is caused by minute variations of direct current flowing in certain types of devices. Direct current must flow in a device in order for shot noise to exist, but thermal noise exists in the absence (or presence) of direct current. Temperature and resistance control the amplitude of thermal noise voltage, but they do not appear in the shot-noise current equation.

Both behave the same as far as frequency and bandwidth are concerned, and one cannot be distinquished from the other by its appearance on an oscilloscope or other electronic instrument. As will be shown, the equation for shot noise in a p-n junction, or diode, bears a striking similarity to the equation for thermal noise voltage in a resistor. Although they originate in different ways, shot noise, thermal noise, or a combination of the two may be handled as one type of voltage waveform in measurement and design activities.

3–4. *How is the value of shot-noise current calculated?*

The value of shot-noise current is defined as:

$$i_{sn} = \sqrt{2eIB} \qquad \text{(Eq. 3-1)}$$

where,

i_{sn} is the rms value of shot-noise current in amperes,
e is the charge of an electron (1.6×10^{-19} coulomb),
I is direct current in amperes,
B is noise bandwidth of the system in hertz.

The bandwidth term, B, in Eq. 3-1 is the same equivalent rectangular bandwidth term used in Eq. 2-1 for thermal noise voltage, and it will be discussed fully in Part 4.

Equation 3-1 may be simplified by substituting the value of e and taking it and the factor 2 outside of the square-root sign.

$$\begin{aligned} i_{sn} &= \sqrt{2eIB} \\ &= 5.65 \times 10^{-10} \sqrt{IB} \qquad \text{(Eq. 3-2)} \end{aligned}$$

Notice the similarity between Eq. 3-2 and Eq. 2-4. The constants are different, and the resistance term in the thermal noise equation has been replaced by direct current in the shot-noise equation. Otherwise, you will find the forms of the two equations are the same.

Equation 3-1 shows that shot-noise current can be minimized by decreasing direct current and bandwidth to their lowest usable values. In some cases, however, other factors may result in a smaller overall shot-noise output if direct current is increased. This is illustrated in Question 3-6, with shot noise in a forward-biased p-n junction (diode) and with avalanche noise in a zener diode.

3–5. *What are typical values of shot-noise current?*

Equation 3-2 indicates that shot-noise current would be infinite if bandwidth could be infinite, but no device can have an infinite bandwidth. If a direct current of 1000 amperes flows through a device having a noise bandwidth of 1000 MHz, the shot-noise current would have a value of only 0.565 milliampere.

A more practical example is 10 milliamperes of direct current flowing in a device with a noise bandwidth of 10 kHz. The shot-noise current in this case would calculate to be:

$$
\begin{aligned}
i_{sn} &= 5.65 \times 10^{-10} \sqrt{IB} \\
&= 5.65 \times 10^{-10} \sqrt{(10^{-2})\,(10^{4})} \\
&= 5.65 \times 10^{-10}\,(10) \\
&= 5.65 \times 10^{-9} \\
&= 0.00565\ \mu A
\end{aligned}
$$

If this current is flowing through a 1000-ohm resistor, the shot-noise voltage produced across the resistor will be:

$$
\begin{aligned}
e_{sn} &= i_{sn}\,(R) \\
&= 0.00565\ \mu A \times 1000\ \text{ohms} \\
&= 5.65\ \mu V
\end{aligned}
$$

Should the value of direct current be reduced from 10 milliamperes to 100 microamperes, the shot noise would be reduced to one-tenth of the value shown above.

3–6. *What is the value of shot-noise voltage in a p-n junction?*

Fig. 3-1 shows the shot-noise equivalent circuit of a forward-biased p-n junction or diode. It consists of constant-current, shot-noise generator i_{sn} in parallel with dynamic resistance r_d,

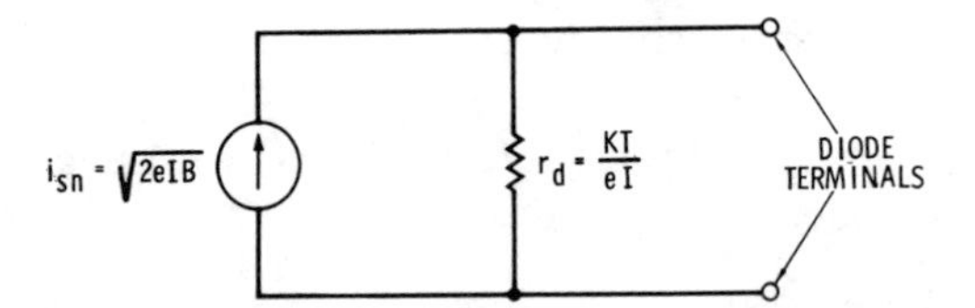

Figf. 3-1. Shot-noise equivalent circuit of a forward-biased p-n junction (diode).

of the p-n junction. It is assumed that an external current source causes direct current I to flow through the diode in the forward direction. This causes the shot-noise generator to generate shot-noise current. It is further assumed that the external direct current source has a series resistance which is very much larger than r_d so that essentially all of the shot-noise current flows through r_d. Under these conditions, the shot-noise voltage appearing across the diode terminals is:

$$e_{sn} = i_{sn} r_d \qquad \text{(Eq. 3-3)}$$

Then, substituting the value of i_{sn} given in Eq. 3-1 in Eq. 3-3 gives:

$$e_{sn} = \sqrt{2eIB}\,(r_d) \qquad \text{(Eq. 3-4)}$$

The value of r_d may be expressed as:

$$r_d = \frac{KT}{eI} \qquad \text{(Eq. 3-5)}$$

where,

r_d is the dynamic resistance of the p-n junction in ohms,
K is Boltzmann's constant (1.38×10^{-23} joules/degree Kelvin),
T is absolute temperature in degrees Kelvin,
e is the charge of an electron (1.6×10^{-19} coulomb),
I is direct current in amperes.

Substituting this value of r_d into Eq. 3-4 gives:

$$e_{sn} = \sqrt{2eIB}\left(\frac{KT}{eI}\right) \qquad \text{(Eq. 3-6)}$$

The quantity in parenthesis, which is r_d, may be squared and placed under the square-root sign:

$$e_{sn} = \sqrt{2eIB\left(\frac{KT}{eI}\right)\left(\frac{KT}{eI}\right)} \qquad \text{(Eq. 3-7)}$$

Simplifying:

$$e_{ns} = \sqrt{2KTB\left(\frac{KT}{eI}\right)} \qquad \text{(Eq. 3-8)}$$

This shows that the value of shot-noise voltage across the diode terminals becomes smaller as direct current increases. Even though the value of shot-noise current increases with higher values of direct current, the dynamic resistance decreases more rapidly, causing a lower value of noise voltage. If r_d is substituted for the quantity in parenthesis in Eq. 3-8, the equation for shot-noise voltage becomes very much like Eq. 2-1, the expression for thermal noise voltage:

$$e_{sn} = \sqrt{2KTB\, r_d} \qquad \text{(Eq. 3-9)}$$

At 290 K, Eq. 3-5 becomes:

$$r_d = \frac{.025}{I} \qquad \text{(Eq. 3-10)}$$

If direct current is expressed in milliamperes:

$$r_d = \frac{25}{I_{(mA)}} \qquad \text{(Eq. 3-11)}$$

This shows that the dynamic resistance of the forward-biased diode is quite low for even moderate values of direct current.

3–7. *What is a temperature-limited diode?*

A temperature-limited diode is a vacuum-tube diode which has a dc plate current that is controlled or limited by the temperature of its filament or cathode. In practice, the filament temperature, and thus the plate current, is controlled by varying the voltage which heats the filament. Fig. 3-2 is a schematic showing how this may be accomplished; a rheostat is used in this circuit to vary filament temperature. A directly heated tungsten filament is commonly used as the cathode because it displays good temperature-limited characteristics.

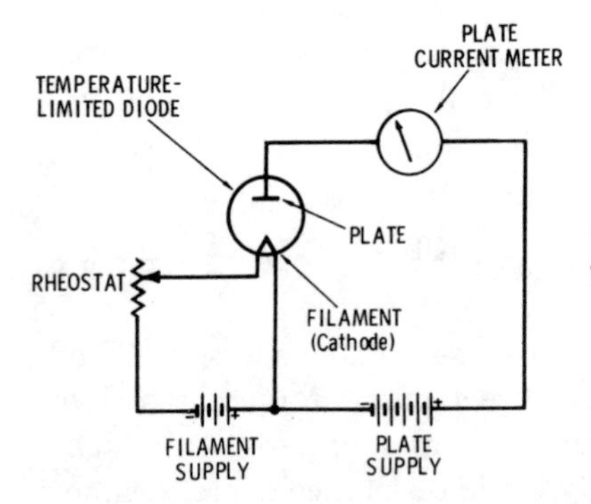

Fig. 3-2. Basic configuration of a temperature-limited diode circuit.

Fig. 3-3 shows how the plate current of a temperature-limited diode changes as plate voltage and filament temperature are varied. At very low plate voltages, the plate current is determined almost entirely by the amplitude of the plate voltage. At higher plate voltages (100 volts or more), the plate current changes much less due to plate voltage variations, but it may be

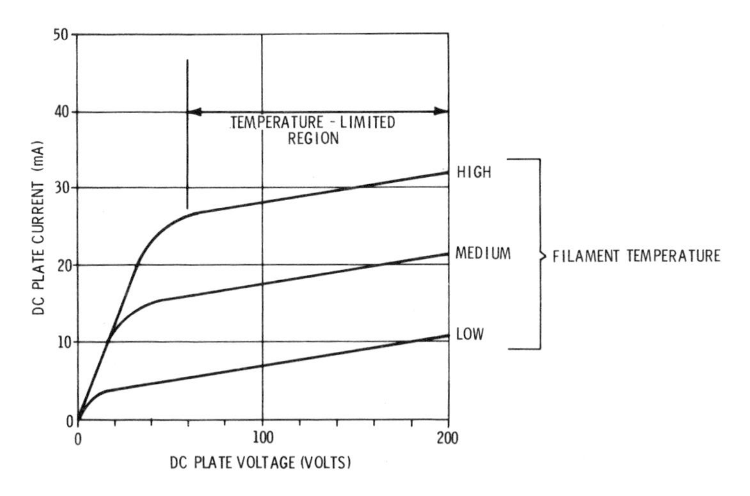

Fig. 3-3. Plate characteristics of a temperature-limited diode.

easily controlled by filament temperature. In the temperature-limited region of Fig. 3-3, the slope of the curves are such that the dynamic plate resistance of the diode is on the order of 10,000 ohms or more. Unlike the semiconductor p-n junction or diode, the temperature-limited diode tends to display a constant value of dynamic resistance as diode current varies.

3–8. *How is a temperature-limited diode useful in the study of noise?*

Shot noise in temperature-limited diodes can be predicted accurately by Eq. 3-1 or Eq. 3-2, where I is the diode plate current. This, plus the fact that its dynamic plate resistance has a relatively high and constant value, makes the temperature-limited diode very useful as a noise generator.

Fig. 3-4 shows the shot-noise equivalent circuit of a temperature-limited diode. If an external load resistor, which is very small compared to 10,000 ohms, is connected across the diode, then essentially all of the shot-noise current will flow through this external load resistor, and the dynamic plate resistance of the diode may be neglected. Fig. 3-5 shows this equivalent circuit. Thus a shot-noise voltage generator is produced which has a predictable output noise voltage and a constant source resist-

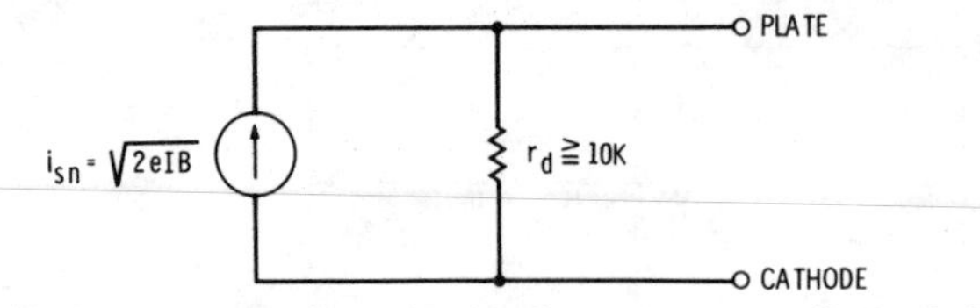

Fig. 3-4. Shot-noise equivalent circuit of a temperature-limited diode.

ance. Within limits, this source resistance may be easily varied. The output shot-noise voltage is the product of shot-noise current i_{sn} and load resistance R_g.

$$\begin{aligned} e_{sn} &= i_{sn} R_g \\ &= R_g \sqrt{2eIB} \end{aligned} \qquad \text{(Eq. 3-12)}$$

The equivalent circuit may also be expressed using a noise voltage generator, as in Fig. 3-6.

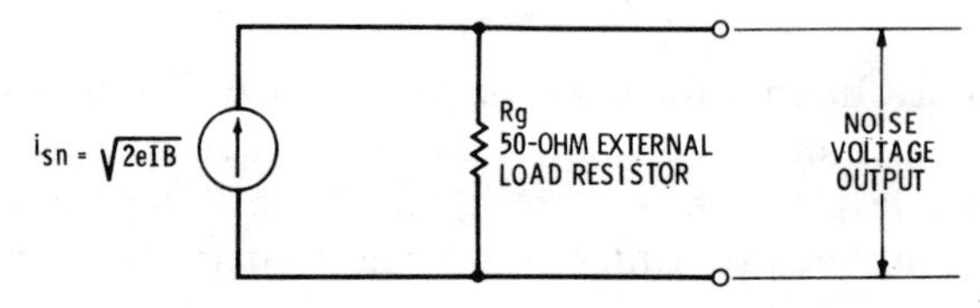

Fig. 3-5. Shot-noise equivalent circuit of a temperature-limited diode with a 50-ohm load resistor.

A temperature-limited diode noise generator is very useful in noise figure measurements, because it eliminates the need to know the noise bandwidth of the system being measured. This feature can save considerable measurement effort and eliminate uncertainties regarding noise bandwidth. The use of the temperature-limited diode noise generator for measuring noise figure is described in Part 7.

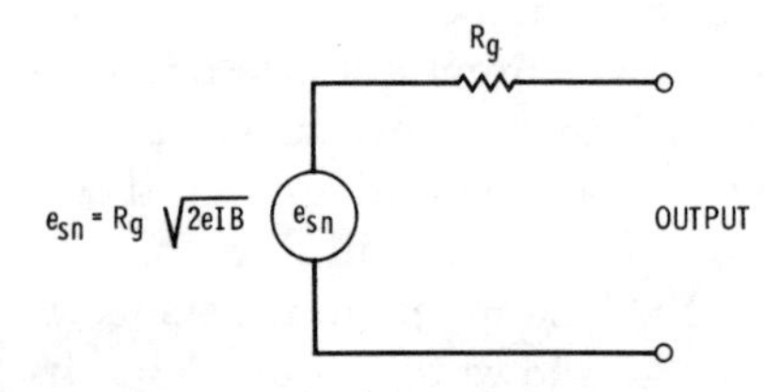

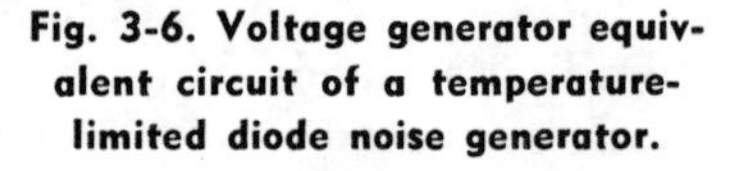

Fig. 3-6. Voltage generator equivalent circuit of a temperature-limited diode noise generator.

3–9. *What are typical values of shot-noise voltage from a temperature-limited diode noise generator?*

Diodes designed for use as shot-noise generators usually have limited plate current capability. The Sylvania 5722 diode has an absolute maximum plate current rating of 35 milliamperes. Table 3-1 shows the shot-noise output voltage for different noise bandwidths from a temperature-limited diode generator having a dc plate current of 30 milliamperes and a source resistance of 50 ohms. Values given are simply solutions to Eq. 3-12.

Table 3-1. Temperature-Limited Diode Noise Generator Voltage Output (Plate current of 30 mA and R_g of 50 ohms)

Noise Bandwidth	Shot-Noise Voltage
500 Hz	0.11 microvolt
5 kHz	0.35 microvolt
50 kHz	1.1 microvolt
500 kHz	3.5 microvolts
5 MHz	11.0 microvolts

3–10. *How may a simple temperature-limited diode noise generator be built?*

Fig. 3-7 shows a schematic diagram of a simple diode noise generator. Filament temperature is adjusted by means of rheo-

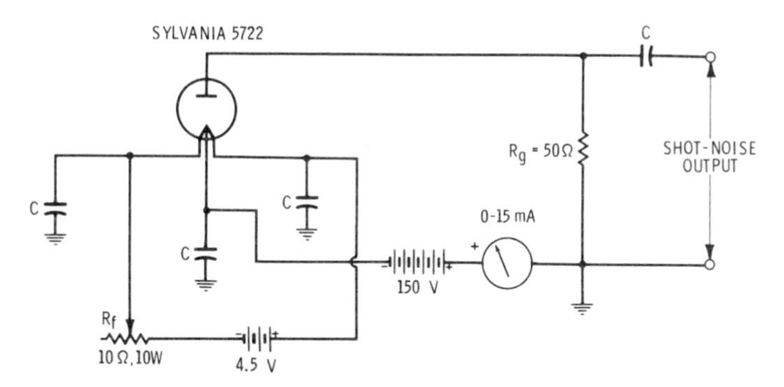

Fig. 3-7. Shot-noise generator using a temperature-limited diode.

stat R_f to obtain a desired dc plate current, as indicated on the milliammeter. The entire generator should be well shielded in a metal box. If ac-operated power supplies are used to furnish the dc-operating voltages, they should be well filtered and bypassed with suitable capacitors for the frequency range of interest. The four capacitors in Fig. 3-7 should have negligible impedance, compared to 50 ohms, at the operating frequency. Other values of R_g may be used, up to several hundred ohms, but the higher values may contribute to noticeable errors in noise output voltage. R_g should have as little reactance as possible at the measurement frequency.

4

Noise Bandwidth

4–1. *What is noise bandwidth?*

Noise bandwidth is a term used in the equations for thermal noise and shot noise. Its value is not necessarily the same as the 3-dB, or half-power, bandwidth commonly specified in various systems. Some of the more sophisticated amplifiers and filters closely approach an ideal rectangular passband, but simpler systems may have noise bandwidths considerably wider than their 3-dB bandwidths.

Fig. 4-1 shows an actual power frequency response (solid curve) and its equivalent rectangular power response (dashed). In this case, the noise bandwidth is significantly larger than the 3-dB bandwidth. The equivalent rectangle must contain the same area as the total area under the solid curve, and it must have the same height as the solid curve. When these two conditions exist, the width of the rectangle in hertz is the noise bandwidth used in calculating thermal noise and shot noise.

Notice that the horizontal and vertical scales of Fig. 4-1 are linear. This is necessary so that linear areas may be determined. The actual frequency response used to determine noise bandwidth must be a power frequency response, rather than a voltage frequency response. Voltage frequency response curves have different shapes than corresponding power response curves, and they should not be used for determining noise bandwidth. A power frequency response curve shows how output power varies versus frequency as a constant-amplitude input signal is swept across the passband.

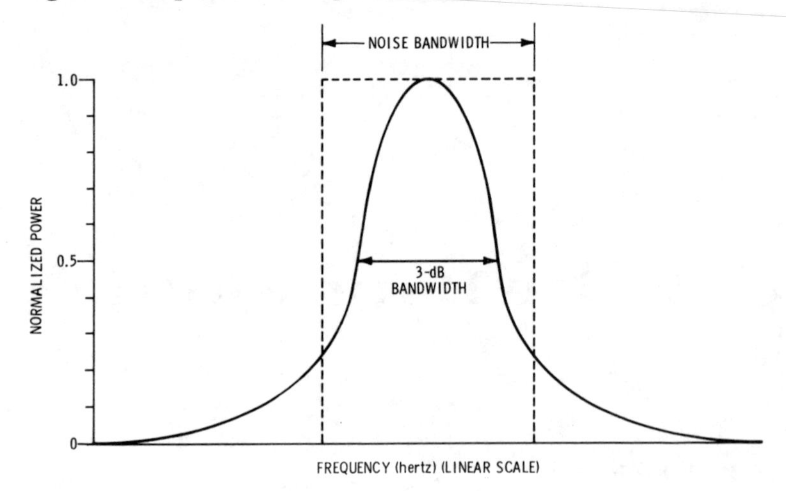

Fig. 4-1. Actual bandpass response (solid) and equivalent rectangular response (dashed).

4-2. *How can noise bandwidth be found graphically?*

The equivalent rectangular noise bandwidth of a system may be determined graphically by taking data on the system, plotting the power response curve, and graphically determining the area under the curve. This area is divided by the height of the curve to find the width (bandwidth) of the equivalent rectangle. The plotted curve must be a power response, rather than a voltage response. Fig. 4-2 shows an equipment arrangement for taking data. The output power of this system will be:

$$P_o = \frac{(E_o)^2}{R} \qquad \text{(Eq. 4-1)}$$

Since R will be constant for all readings of data, output power may be taken as proportional to the square of output voltage E_o. Signal generator output voltage is held constant at some convenient value as its frequency is adjusted to various discrete values at which output voltages are measured and recorded.

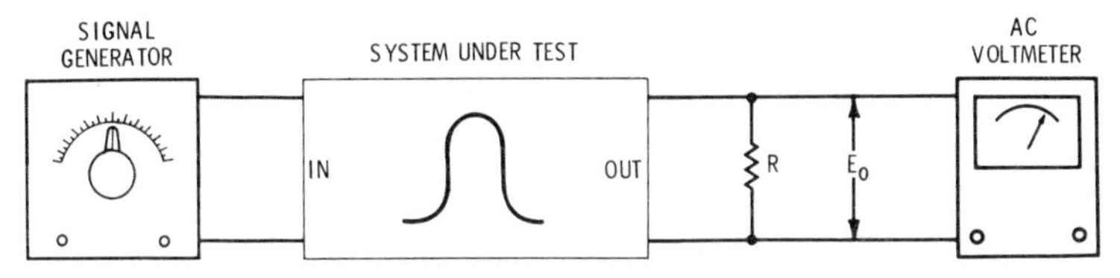

Fig. 4-2. Test arrangement for measuring power frequency response.

The number of these discrete frequencies should be sufficient to describe the passband of the system. Care must be taken that the output signal level is large compared to output noise at mid-band, but not so large that clipping of the signal occurs. Output voltage at each discrete frequency is recorded; then, each value of output voltage is squared, giving quantities proportional to output power.

In finding noise bandwidth, it is convenient to plot the curve as normalized power, such that the peak response has a value of 1.0 and all other points have values less than unity. The squared output voltages may be normalized by dividing each by the highest recorded value of squared output voltage. The curve may then be plotted on linear graph paper (do not use semilog or log-log paper) as frequency versus normalized, squared output voltage. This will result in the same curve as a normalized power curve.

The area under the curve may be found by counting the graph paper grid squares under the curve, keeping track of fractions of squares where the curve crosses them. This area is divided by the maximum height of the curve (in squares) to find the width (in squares) of the equivalent rectangle. Referring this number of squares to the horizontal-frequency axis will yield the equivalent rectangular noise bandwidth in hertz. Instead of counting squares, a planimeter may be used to find the area under the curve in square inches. This area is then divided by the maximum height of the curve in inches to find the noise bandwidth in inches. Inches may then be converted to hertz by laying a ruler along the frequency axis of the graph.

The graphical method of finding noise bandwidth is valuable when irregular shaped passbands are encountered. In many instances, however, the degree of accuracy required does not warrant this much effort. If the circuit parameters which govern the shape of the passband are known, the noise bandwidth may be calculated and related to the 3-dB bandwidth.

4-3. *How can noise bandwidth be calculated?*

Computation of the area under a power frequency response curve requires the use of complex algebra and integral calculus; the mathematics become rather involved as the number of reactive elements in the circuit increases.

First, the mathematical transfer function of the frequency-selective network is written. This equation expresses the ratio of output voltage to input voltage as a function of frequency. All voltage amplifiers are considered to have a voltage gain of one, and the maximum value the transfer function may have is unity. This normalizes the response curve as in the graphical method. The transfer function is then converted to polar form and the phase angle, which has no effect on bandwidth, is ignored. The polar form of the equation is squared to give the power transfer function, and this function is integrated with respect to frequency from zero to infinity. The result is the area under the normalized power curve. Since the maximum height of the curve is unity, the numerical value of the rectangular bandwidth is equal to the value of the area calculated. Dividing rectangular bandwidth by the 3-dB bandwidth yields a convenient factor which may be used to convert 3-dB bandwidth to noise bandwidth.

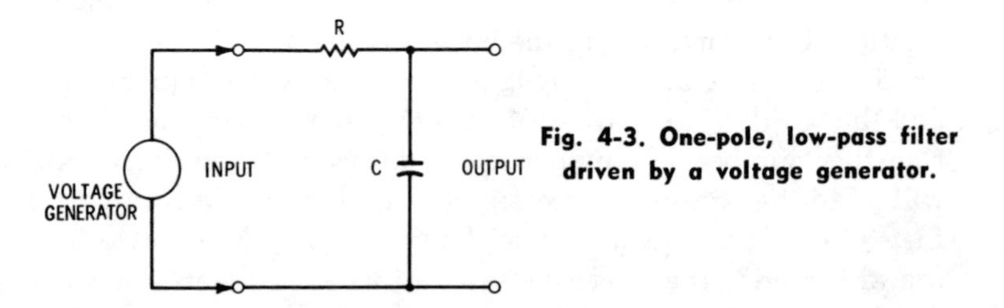

Fig. 4-3. One-pole, low-pass filter driven by a voltage generator.

The simplest example is a one-pole, low-pass filter, such as shown in Fig. 4-3. The 3-dB bandwidth is:

$$B_{3dB} = \frac{1}{2\pi RC} \qquad \text{(Eq. 4-2)}$$

where,

B is in hertz,
R is in ohms,
C is in farads.

High-frequency response falls at the rate of 6 dB per octave. Performing the mathematical operations described above:

$$T_f = \frac{1}{1 + j(2\pi fRC)}$$

$$= \frac{1}{\sqrt{1 + (2\pi fRC)^2}} \qquad \text{(Eq. 4-3)}$$

where,

T_f is the voltage transfer function,
f is frequency in hertz.

Squaring the transfer function and integrating result in the rectangular noise bandwidth.

$$B_n = \int_0^\infty \frac{df}{1 + (2\pi fRC)^2}$$

$$= \frac{1}{4RC} \qquad \text{(Eq. 4-4)}$$

Dividing the noise bandwidth by the 3-dB bandwidth yields:

$$\frac{B_n}{B_{3dB}} = \frac{\frac{1}{4RC}}{\frac{1}{2\pi RC}}$$

$$= \frac{\pi}{2}$$

$$= 1.57 \qquad \text{(Eq. 4-5)}$$

The 3-dB bandwidth is easily calculated or measured and, with Eq. 4-5, it is a simple matter to find the noise bandwidth. Fig. 4-4 shows a graph of the normalized power response curve for a low-pass filter, such as shown in Fig. 4-3. Values of R and C are selected such that the 3-dB bandwidth is 3 kHz. From Eq. 4-5, the noise bandwidth is 1.57 times 3 kHz, or 4.71 kHz.

4-4. *How does noise bandwidth typically relate to 3-dB bandwidth?*

The following tables may be used to convert 3-dB bandwidth to equivalent rectangular noise bandwidth, or simply, noise bandwidth. If the 3-dB bandwidth of a system is not known, it may be measured by using the arrangement shown in Fig. 4-2

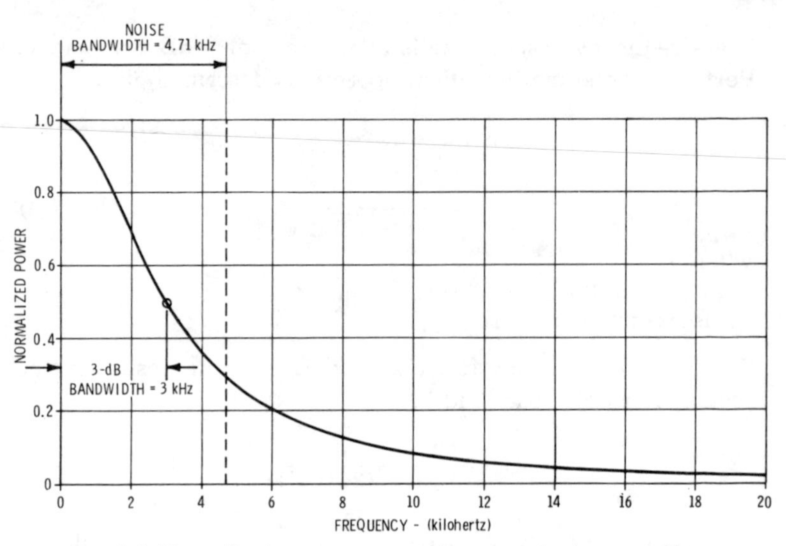

Fig. 4-4. Normalized power response of a one-pole, low-pass filter.

and finding the two frequencies at which output voltage E_o is 0.707 of the maximum or peak output voltage. The difference between these two frequencies is the 3-dB bandwidth.

In order to use the tables, the circuit configuration of the system must be known to the extent of how many reactive elements or resonators are used to control the system bandwidth. The circuit of Fig. 4-5 is a single-stage, wide-band amplifier. The low-frequency, 3-dB point is determined by C_L and R_g, and the high-frequency, 3-dB point is determined by C_H and R_D. Fig. 4-6 shows one stage of a single-tuned, narrow-band amplifier. The bandwidth of this amplifier is determined by the Q of the LC resonator in the drain circuit of the FET. If identical

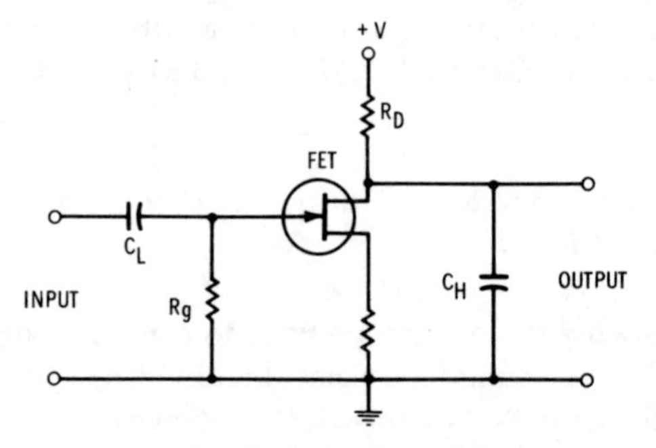

Fig. 4-5. Wide-band amplifier.

stages of wide-band amplifiers are connected in cascade, the noise bandwidth will approach the value of the 3-dB bandwidth as more stages are added. The same is true of the single-tuned narrow-band amplifier. Table 4-1 gives the ratio of noise bandwidth B_n to 3-dB bandwidth B_{3dB} for up to five identical stages

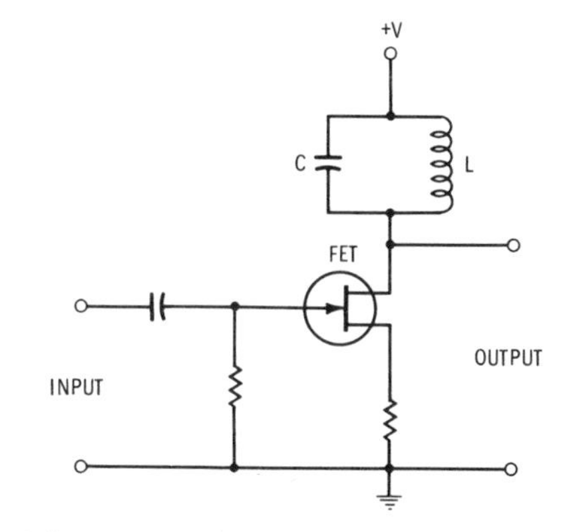

Fig. 4-6. Single-tuned amplifier.

of these two types of circuit. The ratios are correct only when all stages are identical and only when the input impedance of each stage has negligible effect on the output impedance of the preceding stage.

Table 4-1. Noise Bandwidth Factors for Wide-Band Amplifiers and Single-Tuned Amplifiers

Number of Stages	B_n/B_{3dB}
1	1.57
2	1.22
3	1.16
4	1.14
5	1.12

It should be recognized that the 3-dB bandwidth will decrease as more stages are added. The factors given in Table 4-1 refer to the overall 3-dB bandwidth of the stages, not to that of a single stage. Suppose the circuit of Fig. 4-5 has a 3-dB bandwidth of 10 kHz. From Table 4-1, the noise bandwidth is 1.57 times the 3-dB bandwidth, or 15.7 kHz. If three stages identical to Fig. 4-5 are cascaded, the overall 3-dB bandwidth will be reduced to 5.1 kHz. From Table 4-1, the noise bandwidth is 1.16 times the 3-dB bandwidth, or 5.92 kHz (1.16×5.1). In like manner, if two stages identical to Fig. 4-6 have an overall bandwidth of 10 kHz, Table 4-1 shows the overall noise bandwidth is 1.22 times the 3-dB bandwidth, or 12.2 kHz.

Fig. 4-7 shows one stage of a double-tuned amplifier. Table 4-2 lists the ratio of B_n/B_{3dB} for double-tuned stages if the two

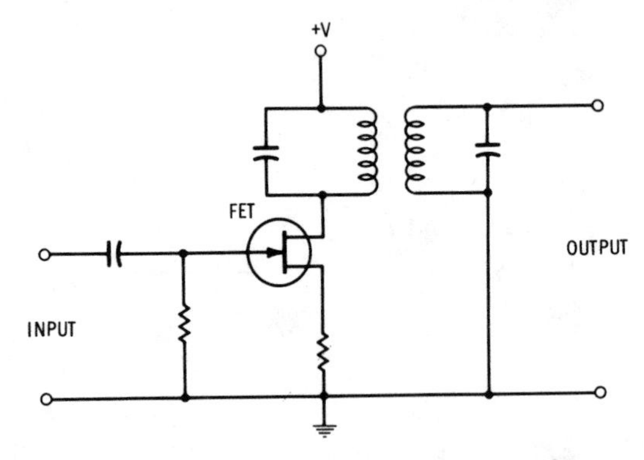

Fig. 4-7. Double-tuned amplifier.

tuned circuits are critically coupled. For three or more identical double-tuned stages, the noise bandwidth may be considered the same as the 3-dB bandwidth.

Table 4-2. Noise Bandwidth Factors for Double-Tuned Amplifiers

Number of Stages	B_n/B_{3dB}
1	1.11
2	1.04

4-5. *How can noise bandwidth be easily controlled?*

Sometimes it is desirable to accurately control the noise bandwidth of a system to a known value. Because of tolerances in component parameters, problems occur in trying to make the frequency-response characteristics of several amplifier stages identical to each other. Interaction between stages may aggravate these problems. One solution, which has been used with success, is to design the amplifier stages so that the overall bandwidth is very much larger than the desired noise bandwidth. The noise bandwidth is then controlled, by means of simple filters, to the desired value. Such filters might take the form of a single-tuned circuit, for a narrow-band application, or simple RC networks, for wide-band circuits. This procedure may not be suitable for some applications, but it should not be overlooked as a possible solution if system constraints allow its use.

5

Special Considerations for Noise

5-1. *Why are special considerations required for noise measurements?*

The characteristics of noise differ markedly from those of direct current or sinusoidal ac waveforms. A noise voltage waveform is quite different from that of a sine wave; there is no coherent phase parameter associated with noise. Bandwidth plays an important role in determining the amplitude and fluctuation of noise.

At the point where noise originates in a circuit, its amplitude is generally quite small, and circuits in which noise is a significant parameter usually include large voltage gains. If extraneous signals get into these circuits, they too may be greatly amplified and cause errors in the noise measurements.

These facts require that some special care and knowledge be applied when working with noise, over and above that provided by basic electronic theory. Noise bandwidth, discussed previously, is a good example of the special considerations required in noise work.

5–2. *What general precautions should be taken in measuring noise?*

The equipment under test should be well shielded, as should the associated test equipment, to prevent unwanted signals or interference from entering the system and causing errors. In audio-frequency systems, 60-Hz hum from power-supply ripple, induced pickup, or ac power-wiring ground loops can present tedious problems. Rf systems may pick up strong signals or interference in spite of elaborate shielding efforts.

An oscilloscope provides a good way to visually monitor the noise being measured. The oscilloscope presentation should resemble the photograph of Fig. 5-1. Any evidence of a repretitive waveform, such as a sine wave or pulse train, indicates that the output contains something more than noise and remedial steps should be taken. Of course, the frequency response of the oscilloscope should be commensurate with the output frequency range of the system. Shielded rooms, called screen rooms, are sometimes necessary to prevent interference from entering equipment under test and destroying the accuracy of noise measurements.

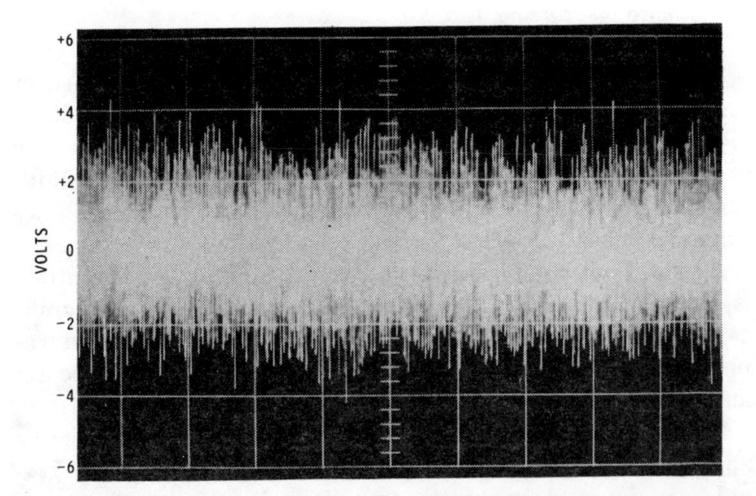

Fig. 5-1. Photograph of noise displayed on an oscilloscope (1.0-volt rms).

5-3. *What is the peak-to-rms ratio of noise?*

It is well known that the peak value of a sine wave is 1.414 times its rms value and the peak-to-peak voltage is 2.828 times the rms voltage. Thus, the peak-to-rms ratio for sine waves is 1.414, and this is a precise, unvarying factor.

Perhaps less well known is the peak-to-rms ratio of noise. Fig. 5-1 is a photograph of noise displayed on an oscilloscope. The bandwidth of this noise is approximately 20 kHz and its amplitude is 1.0 volt true rms. The horizontal scale is 1 millisecond per division and the vertical scale is 2 volts per division. An inspection of the photograph quickly shows that there is no one value of peak amplitude for noise. Different peaks have different amplitudes. Although the amplitudes of the peaks vary in a random manner, some arbitrary level may be chosen which will include most of the peaks. Nearly all of the peaks in the photograph fall within the middle four divisions of the vertical scale, indicating a maximum peak-to-peak amplitude of 8 volts, or a peak amplitude of 4 volts. A very few peaks are in the range of 4 to 5 volts from the horizontal axis (zero volts line), but if they were clipped at the 4-volt level, no significant change would occur in the rms value of the noise voltage. This indicates that a practical value for the maximum peak-to-rms ratio of noise is 4.0. Larger values, such as 5.0, may be used, but they are not required for ordinary noise work.

5-4. *Why is the peak-to-rms ratio of noise important?*

Some circuits, such as amplifiers in noise generators and noise-measuring equipment, must amplify noise with good fidelity. The peak-to-rms ratio of noise is important in these applications. Electronic ac voltmeters used to measure noise should be able to accept peak voltages 4 to 5 times greater than the rms meter reading without clipping the waveform. This is considerably greater than the factor of 1.414 required for sine waves. An amplifier which must have a noise output of 1.0 volt rms should be able to produce an undistorted peak-to-peak output signal of at least 8 volts to preclude significant clipping of the noise signal.

5–5. *How can noise and signal voltages be added?*

Sometimes the sum of two noise voltages, or the sum of a sine-wave voltage and a noise voltage, must be calculated. Since noise is an ac voltage having no coherent phase or repetitive waveform, noise voltages cannot be simply added together as can dc or in-phase, sine-wave voltages.

Fig. 5-2 shows two noise generators connected in series driving a resistive load. The power dissipated in the load resistor R due to each noise generator is:

$$P_1 = \frac{(e_1)^2}{R} \qquad \text{(Eq. 5-1)}$$

$$P_2 = \frac{(e_2)^2}{R} \qquad \text{(Eq. 5-2)}$$

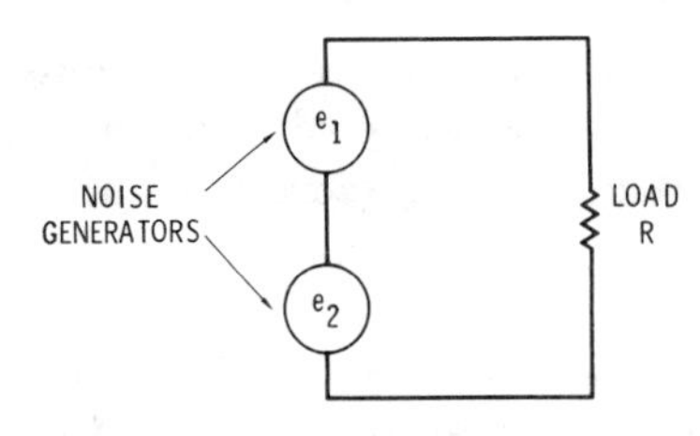

Fig. 5-2. Two noise generators connected in series with load resistor R.

The total power dissipated in the load is the sum of these two powers, or:

$$\begin{aligned} P_T &= P_1 + P_2 \\ &= \frac{(e_1)^2}{R} + \frac{(e_2)^2}{R} \\ &= \frac{(e_1)^2 + (e_2)^2}{R} \end{aligned} \qquad \text{(Eq. 5-3)}$$

An equivalent total noise voltage, e_t, which would dissipate this same total power in the load, could be expressed as:

$$(e_t)^2 = (e_1)^2 + (e_2)^2 \qquad \text{(Eq. 5-4)}$$

$$e_t = \sqrt{(e_1)^2 + (e_2)^2} \qquad \text{(Eq. 5-5)}$$

Equation 5-5 is also used to find the effective sum of a sine-wave voltage and a noise voltage.

If we assume e_1 in Fig. 5-2 to be a noise voltage of 3 microvolts and assume e_2 to be 4 microvolts, then from Eq. 5-5, the equivalent total voltage is:

$$\begin{aligned} e_t &= \sqrt{(3)^2 + (4)^2} \\ &= \sqrt{9 + 16} \\ &= \sqrt{25} \\ &= 5\ \mu V \qquad \text{(Eq. 5-6)} \end{aligned}$$

Simply adding the two voltages would result in an incorrect answer of 7 microvolts.

5–6. *How can noise and signal voltages be subtracted?*

If the sum of two noise voltages, or the sum of a signal voltage and a noise voltage, e_t, is known, and one of the two, e_1, is also known, the other voltage, e_2, may be found by use of Eq. 5-4 as follows:

$$(e_2)^2 = (e_t)^2 - (e_1)^2 \qquad \text{(Eq. 5-7)}$$

$$e_2 = \sqrt{(e_t)^2 - (e_1)^2} \qquad \text{(Eq. 5-8)}$$

One of the important applications of this expression occurs when it is necessary to find the value of signal output from a system when it is only possible to measure the quantities of noise output and an output which is a combination of signal and noise. Such a combination is commonly referred to as signal plus noise or $(S + N)$. When this occurs, Eq. 5-8 can be rewritten as:

$$S = \sqrt{(S + N)^2 - (N)^2} \qquad \text{(Eq. 5-9)}$$

where,

S is the unknown signal output voltage,
N is the measured output noise voltage when there is no signal input,
$(S + N)$ is the measured output which is a combination of signal and noise voltages.

If the output noise voltage of a system is 3 volts and the combination of signal and noise output voltage is 5 volts, the value of output signal voltage may be found as:

$$
\begin{aligned}
S &= \sqrt{(5)^2 - (3)^2} \\
&= \sqrt{25 - 9} \\
&= \sqrt{16} \\
&= 4 \text{ volts} \qquad \text{(Eq. 5-10)}
\end{aligned}
$$

Simply subtracting the 3 volts of noise voltage from the 5 volts of (S + N) would result in an incorrect value for the signal of 2 volts.

5–7. *What is a true rms ac voltmeter?*

A true rms ac voltmeter is one which measures the true root-mean-square (abbreviated rms) value of a voltage, regardless of the waveform of that voltage. It does not matter whether the applied waveform is a sine wave, a combination of sine waves, a square wave, a sawtooth wave, a triangle wave, a noise waveform, or a combination of waveforms; this type of ac voltmeter will read the correct rms voltage.

The term rms is used to indicate that value of ac voltage which will produce the same amount of heat in a resistor as will an equal value of dc voltage. For example, 10 volts dc applied to a 5-ohm resistor will cause a power dissipation of 20 watts in the resistor. The value of ac rms voltage, regardless of waveform, which will also produce 20 watts of dissipation in the same resistor is 10 volts.

True rms ac voltmeters, which are usually labeled as such, tend to be more expensive than the average-reading type of ac voltmeter, consequently they are not seen as widely in electronics laboratories as are the latter type. Since true rms voltmeters are the most accurate for noise measurements, they are preferable for all noise work.

5–8. *What is an average-reading ac voltmeter?*

An average-reading ac voltmeter is one in which the ac waveform is rectified by one or more diodes; the average dc value of the resulting waveform is indicated by a dc meter. Hence, we have the term, average-reading voltmeter. Most of the less expensive ac voltmeters are of this type. Because they are intended to be used for the measurement of sine-wave voltages, they are calibrated to indicate rms voltage when the ac waveform is a sine wave. Average-reading ac voltmeters range from the simple vom, having only diodes for ac rectification, to electronic voltmeters, in which the diode dc meter circuit is placed in the feedback network of an amplifier to minimize diode nonlinearity and to achieve higher sensitivity.

Different types of ac waveforms have different average dc values when they are rectified. This causes an average-reading voltmeter to respond differently to different waveforms which have the same value of rms voltage. A significant error will occur when noise voltage is measured using an average-reading ac voltmeter.

5-9. *What correction factor should be applied when using an average-reading ac voltmeter to measure noise?*

It can be shown that if the same noise voltage is fed to a true rms ac voltmeter and to an average-reading ac voltmeter, the true rms voltmeter will read 1.05-dB higher than the average-reading voltmeter. This means that noise voltage measured on an average-reading voltmeter should be multiplied by 1.13 to find the true rms noise voltage.

This writer has verified the correction factor of 1.13 by the test illustrated in Fig. 5-3. Both simple voms and electronic ac voltmeters were used as the average-reading voltmeter with no detectable difference in results. It is important to remember

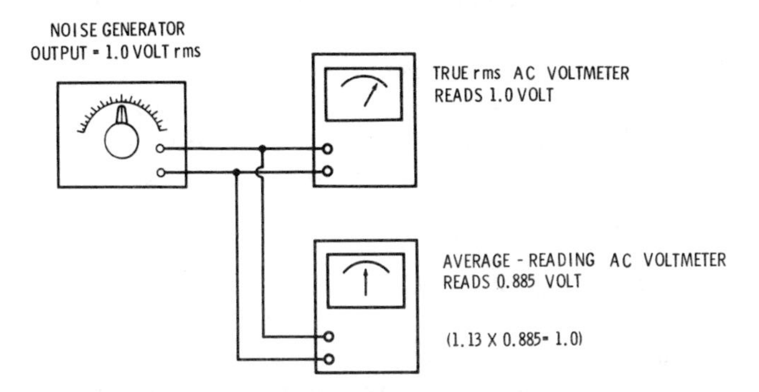

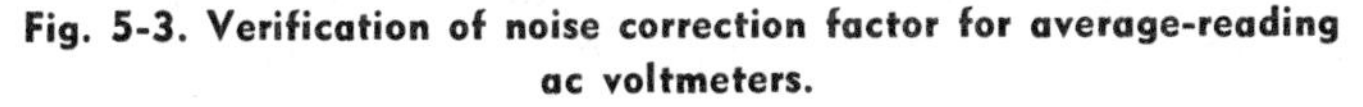
Fig. 5-3. Verification of noise correction factor for average-reading ac voltmeters.

that this correction factor is valid only when the measured waveform is pure noise which contains no other significant component.

5–10. *When may an average-reading ac voltmeter be used for noise measurements?*

Average-reading ac voltmeters may be used to measure pure noise voltages by applying the previously discussed correction factor of 1.13. When the ratio of two measured noise voltages must be calculated, however, the correction factor may be neglected because it would apply to both readings, thus cancelling itself out of the ratio fraction. One noise figure measurement technique requires that a noise generator be used to increase the system noise output by a factor of 1.414 (3-dB voltage ratio). This may be done when using an average-reading voltmeter, without correcting the readings to true rms values. The ratio of the readings will be the same whether or not the correction factor is applied.

5–11. *When should an avearge-reading ac voltmeter not be used in noise measurements?*

When voltages which are a combination of noise and a sine wave are measured on an average-reading ac voltmeter, the correction factor required for finding the true rms value will depend on the relative amplitudes of the noise voltage and the sine-wave voltage. Pure noise would require a correction factor of 1.13; a sine wave would require no correction factor at all, since its correction factor would be 1.00. For best accuracy, a true rms voltmeter should be used to measure voltages which are a combination of noise and other components.

5–12. *What is voltmeter time constant?*

The time constant of a voltmeter is a parameter which indicates how much time is required for the meter to respond to a changing voltage. If a dc voltmeter is connected to a battery, the pointer of the meter will move from a reading of zero to a reading which is 63% of the total battery voltage in an amount of time equal to the time constant of that meter. A time equal to about five time constants is required for the meter to reach its steady-state reading. Suppose a dc voltmeter having a full-scale calibration of 10 volts is connected to a 10-volt battery. If the time constant of the meter is 0.1 second, then the pointer of the meter will move from zero to 6.3 volts in 0.1 second and to a reading of 10 volts in about 0.5 second.

Time constant is very important in ac voltmeters. If a rectified 60-Hz sine wave could be applied to a hypothetical dc meter having a time constant of zero seconds, the pointer of such a meter would move back and forth from zero to the peak value of the sine wave so rapidly that it would be impossible to take any meaningful reading. An actual ac meter, however, has a time constant which is large compared to the period of a 60-Hz sine wave, and it shows a steady, constant reading. Thus, the time constant filters out the voltage fluctuations and shows an averaged reading.

5-13. *Why is voltmeter time constant important in noise measurements?*

The random nature of noise causes fluctuations of its amplitude from instant to instant, and these fluctuations produce random motions of the pointer on a voltmeter used to measure noise. If the amplitudes of the pointer fluctuations are small compared to the average value of the noise voltage (only a few percent), the pointer may be observed for a few seconds and a reasonably accurate average reading can be estimated. This is called time averaging. If the amplitude of the fluctuations are large, the pointer makes excursions covering a large portion of the meter scale in a random manner, and an accurate reading is not feasible.

In order to obtain a reasonably stable noise reading from a voltmeter, its time constant must bear a certain relationship to the bandwidth of the noise being measured. Longer meter time constants are required as the noise bandwidth is decreased. Most meter time constants are sufficiently large for noise measurements if the noise bandwidth is a few kilohertz or larger.

When very narrow noise bandwidths are encountered, the meter time constant must be increased by some means.

5-14. *How can required voltmeter time constant be calculated?*

An expression relating maximum fluctuation to noise bandwidth and meter time constant is given in Eq. 5-11. This relationship has proved to be practical and accurate in the laboratory:

$$\Delta = \frac{1}{\sqrt{2Bt}} \qquad \text{(Eq. 5-11)}$$

where,

Δ is the maximum fluctuation error,
B is noise bandwidth in hertz,
t is meter time constant in seconds.

Maximum fluctuation error Δ is the ratio of maximum fluctuation amplitude to the actual average value of the noise reading. For example, if the actual average value of the noise being measured is 100 millivolts, but the pointer of the meter fluctuates away from this value by as much as 15 millivolts, then Δ would have a value of:

$$\Delta = \frac{15}{100}$$
$$= 0.15$$

In the usual situation, noise bandwidth is known, and an acceptable fluctuation error can be chosen. It is then necessary to calculate the meter time constant required to obtain the desired fluctuation error. Solving Eq. 5-11 for t results in:

$$t = \frac{1}{2B\Delta^2} \qquad \text{(Eq. 5-12)}$$

Reasonably accurate meter readings can be made by time averaging if Δ is about 0.05 or less. The meter time constant required for a maximum fluctuation error of 5% (0.05) and a noise bandwidth of 50 Hz is:

$$t = \frac{1}{2(50)(.05)^2}$$
$$= 4 \text{ seconds}$$

5–15. *What are typical values of meter time constant required in noise measurement?*

Table 5-1 shows the solutions to Eq. 5-12 for several values of noise bandwidth and maximum fluctuation error. Meter time constant, t, may be read at the intersection of the row corresponding to bandwidth, B, and the column corresponding to fluctuation error, Δ.

Table 5-1. Meter Time Constants (in Seconds) for Various Noise Bandwidths and Fluctuation Errors

B	Δ = 1%	Δ = 5%	Δ = 10%
1.0 Hz	5000 s	200 s	50 s
10 Hz	500 s	20 s	5 s
100 Hz	50 s	2 s	0.5 s
1000 Hz	5 s	0.2 s	0.05 s
10,000 Hz	0.5 s	0.02 s	0.005 s

5–16. *How can meter time constant be increased?*

Some types of true rms voltmeters and special noise-measuring equipment have selector switches for controlling meter time constant. If longer time constants are required, it may be possible to modify voltmeters by adding capacitors at certain points in their circuits. Another possibility is to install a more sensitive meter movement with a low-resistance shunt. Such modifications should not be attempted without consulting the voltmeter manufacturer to be certain that such modifications do not alter other voltmeter characteristics, such as accuracy and frequency response. In some situations, it may prove worthwhile to design and build a special-purpose long time constant voltmeter.

5–17. *How can a simple long time constant voltmeter be built?*

A simple average-reading type of ac voltmeter can be built as shown in Fig. 5-4. This circuit has a provision for long time constants by adjusting the values of resistor R and capacitor C. Three integrated circuit operational amplifiers are used. Linear

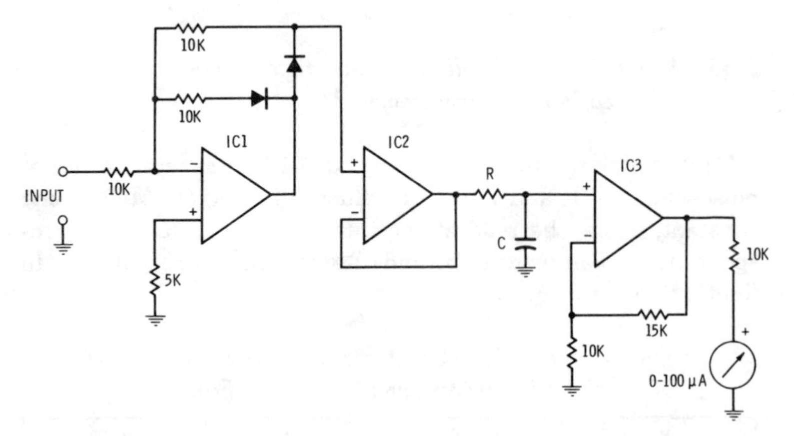

Fig. 5-4. Ac voltmeter circuit with provision for long time constant.

rectification of the ac waveform is performed by IC1, and IC2 is a buffer voltage follower which provides the RC time constant circuit with a constant source resistance. Resistor R and capacitor C are selected to give the desired meter time constant. The dc voltage is amplified by IC3, so that a 1.0-volt true rms noise input voltage will cause the meter to read full scale. The meter time constant is given by:

$$t = RC \qquad \text{(Eq 5-13)}$$

where,

t is time constant in seconds,
R is resistance in ohms,
C is capacitance in farads.

If R is chosen to be 100K, then the value of C for a particular time constant would be:

$$C = \frac{t}{100{,}000} \qquad \text{(Eq. 5-14)}$$

If the required meter time constant is 10 seconds, then:

$$C = \frac{10}{100{,}000}$$
$$= 10^{-4} \text{ farads}$$
$$= 100\ \mu F \qquad \text{(Eq. 5-15)}$$

In designing circuits of this type, care should be taken to observe a peak-to-rms ratio of at least 4.0 for noise waveforms.

5-18. *Where in a system should noise bandwidth be controlled?*

Ideally, the filter which determines the noise bandwidth observed should be located at the output of the system, just ahead of the ac voltmeter used to measure output noise voltage. The reason for this is illustrated in Fig. 5-5. The system under test has a noise bandwidth of 100 Hz, and its output noise voltage is on the order of 2 microvolts. To bring this output noise voltage up to a level more easily measured, a postamplifier is used between the system output and the ac voltmeter. The bandwidth of the postamplifier is 1.0 MHz, and its equivalent input noise voltage (due to its own internal noise) is 5 microvolts. Such a level of equivalent input noise makes it impossible to accurately measure the 2-microvolt output noise from the system under test. If a 100-Hz bandpass filter (same center frequency as the system under test) is inserted between the postamplifier and the voltmeter, however, the equivalent input noise voltage of the postamplifier will be reduced to 0.05 microvolt. This level is negligible compared to the 2-microvolt level to be measured.

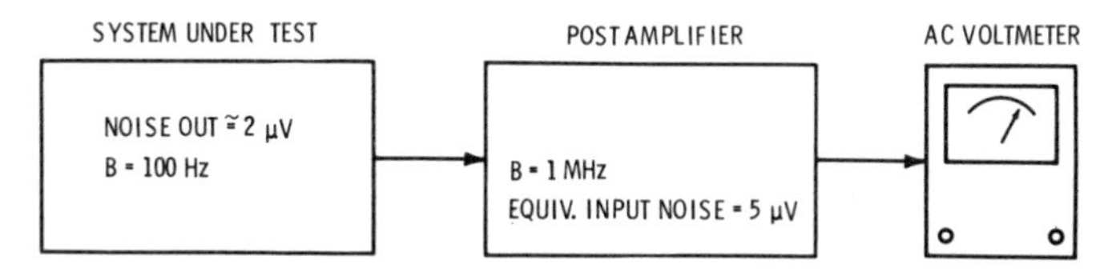

Fig. 5-5. Using a postamplifier in a test system.

Noise voltage varies as the square root of the bandwidth. In the above example, bandwidth was reduced from 1.0 MHz to 100 Hz, which is a ratio of 10,000. The square root of 10,000 is 100; so the noise voltage of the postamplifier was reduced from 5 microvolts to 0.05 microvolt, a reduction of 100 to 1.

6

Signal-to-Noise Ratio

6–1. *What is signal-to-noise ratio?*

Signal-to-noise ratio is defined as the ratio of signal power to noise power existing at some specified point in an electronic system. Usually, the signal-to-noise ratio is of most interest at the input and at the output of a system. It cannot be stated as a ratio of signal voltage to noise voltage, unless the voltage ratio is expressed in decibels. Since signal-to-noise ratio (abbreviated snr) is usually expressed in decibels (dB), voltage ratios are often used to arrive at snr in dB.

6–2. *Why is signal-to-noise ratio important?*

The signal is the desired portion of a system output and the noise is the undesired portion of the output; therefore, the value of signal-to-noise ratio should be as large as possible to ensure a high-quality output signal. Different types of signals can tolerate different levels of noise and still be usable. Morse code can be used for communication at very low values of signal-to-noise ratio, while these same values would render a television signal useless. To operation properly, a system must have some minimum value of output signal-to-noise ratio. Once the value for this is established, receiver noise figure and system losses may be used to determine the antenna gain and transmitter power required to achieve the desired output signal-to-noise ratio.

6–3. *How is signal-to-noise ratio expressed mathematically?*

Equation 6-1 gives the expression for signal-to-noise ratio.

$$\text{snr} = \frac{P_s}{P_n} = \frac{\frac{(e_s)^2}{R}}{\frac{(e_n)^2}{R}} = \frac{(e_s)^2}{(e_n)^2} = \left(\frac{e_s}{e_n}\right)^2 \qquad \text{(Eq. 6-1)}$$

where,

P is power,
e is voltage,
subscripts s and n are used to indicate signal and noise quantities respectively,
R is the resistance into which the signal and noise voltages dissipate power.

Since R is the same for both signal and noise, it cancels out of the equation, leaving the square of a voltage ratio.

Fig. 6-1 shows an equivalent circuit of a signal source with internal resistance R_g and having separate internal signal and

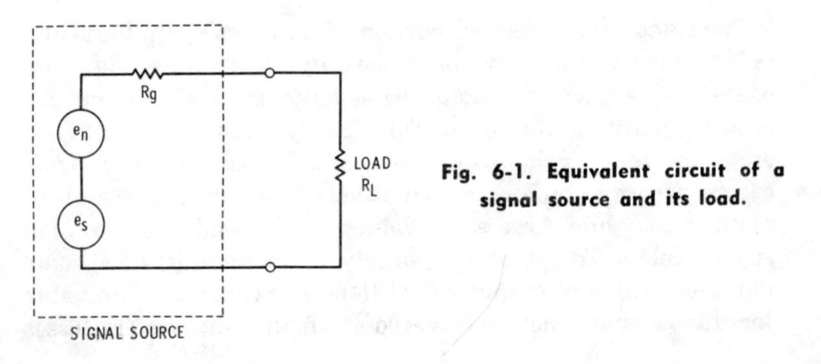

Fig. 6-1. Equivalent circuit of a signal source and its load.

noise generators. Load resistance R_L is connected to the terminals of the signal source. The signal source could be an antenna, a microphone, or a signal generator; the important fact is that it has a signal output and a noise output. Equation 6-1 shows that the signal-to-noise ratio is independent of circuit resistance and is equal to the square of the internal signal voltage divided by the square of the internal noise voltage. These are the voltages that appear at the terminals of the signal source when those terminals are open circuited. Thus, Eq. 6-1 expresses the signal-to-noise ratio of a signal source when no load is applied to its terminals. This is the maximum value of signal-to-noise ratio that the signal source can have. It will be shown that the signal-to-noise ratio across load resistor R_L is less than the value given by Eq. 6-1 when noise voltage e_n is the thermal noise of source resistance R_g.

If signal generators are used as signal sources in noise-related work, care should be taken in calculating signal-to-noise ratio. Many signal generators are calibrated to indicate output signal voltage when loaded by a resistance equal to the internal source resistance of the generator. The open-circuit signal voltage of such generators is twice the calibrated value.

As an example of how signal-to-noise ratio can be calculated, consider a signal generator having an internal source resistance of 50 ohms. This generator is to be used in testing a receiver which has a noise bandwidth of 2 MHz. From Eq. 2-4, the thermal noise of the 50-ohm source resistance is:

$$\begin{aligned} e_n &= 1.27 \times 10^{-10} \sqrt{BR} \\ &= 1.27 \times 10^{-10} \sqrt{(2 \times 10^6)\,(50)} \\ &= 1.27\ \mu V \end{aligned} \qquad \text{(Eq. 6-2)}$$

If the open-circuit signal voltage of the generator is adjusted to a value of 12.7 microvolts, then, from Eq. 6-1, the signal-to-noise ratio of this signal source would be:

$$\begin{aligned} snr &= \left(\frac{e_s}{e_n}\right)^2 \\ &= \left(\frac{12.7}{1.27}\right)^2 \\ &= (10)^2 \\ &= 100 \end{aligned} \qquad \text{(Eq. 6-3)}$$

If the signal generator is calibrated to read signal voltage output when loaded with 50 ohms, it should be set at 6.35 microvolts to obtain this signal-to-noise ratio of 100. Equation 6-3 gives the maximum possible value of input signal-to-noise ratio to the receiver for that signal voltage, source resistance, and noise bandwidth.

6–4. *How is signal-to-noise ratio expressed in decibels?*

Equation 6-4 shows how to convert a signal-to-noise power ratio to decibels (dB), and Eq. 6-5 shows how to convert a signal-to-noise voltage ratio.

$$dB = 10 \log_{10}\left(\frac{P_s}{P_n}\right) \qquad \text{(Eq. 6-4)}$$

$$dB = 20 \log_{10}\left(\frac{e_s}{e_n}\right) \qquad \text{(Eq. 6-5)}$$

Table 6-1 lists dB for several values of power ratio and voltage ratio. Only ratios of one or larger are shown. Notice that power ratios are the square of the corresponding voltage ratios. From Table 6-1 it is found that the signal-to-noise ratio of Eq. 6-3 is equal to 20 dB, because the ratio of signal voltage to noise voltage is equal to 10.

Table 6-1. Conversion of Power and Voltage Ratio to Decibels (dB)

Decibels (dB)	Power Ratio	Voltage Ratio
0	1	1
3	2	1.414
6	4	2
10	10	3.16
12	16	4
20	100	10
30	1000	31.6
40	10,000	100

6–5. *What is sensitivity?*

The term sensitivity is used in different ways to indicate the minimum input signal to which an electronic system will respond. Oscilloscopes may be specified to have a maximum vertical sensitivity of a given number of millivolts or microvolts per scale division. When used in this way, sensitivity merely implies the gain or amplification of the system; noise is not considered because it is very small compared to the minimum usable signal.

Radio receivers may have a sensitivity specification which states that a certain value of input signal will produce a certain output power in watts, or that it will produce a certain output signal-to-noise condition. A good communications receiver might have a sensitivity specification stated as, "less than 0.25 microvolt for 10-dB signal-plus-noise to noise." This means that an input signal at the antenna terminals of the receiver of less than 0.25 microvolt will result in an output $(S + N)/N$ of 10 dB. Signal generators are used to test receivers to this specification; the noise output is due to the thermal noise of the generator source resistance and the circuit noise in the receiver. Since the amplitude of the noise output depends on the bandwidth of the receiver, this type of sensitivity rating should only be used to compare receiver sensitivity ratings when the bandwidths of the receivers are equal. As the bandwidth (selectivity) of the receiver is reduced, the output noise will decrease and a smaller value of input signal voltage will produce the specified output $(S + N)/N$. Output $(S + N)/N$ is easier to measure than output snr and is a more practical quantity.

Fig. 6-2 shows a typical arrangement for testing receivers to the type of sensitivity specification just described. The receiver

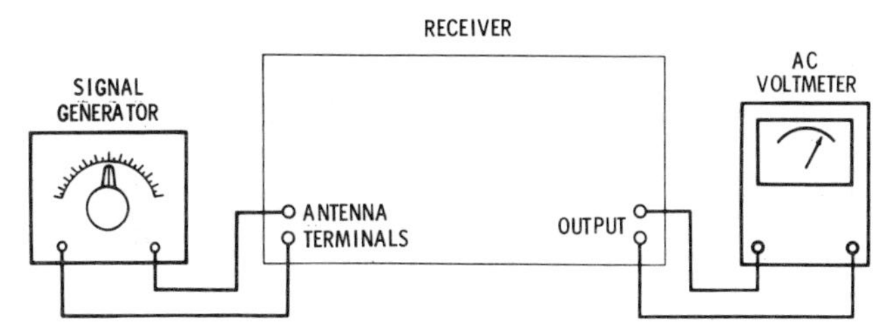

Fig. 6-2. Test arrangement for receiver sensitivity.

and signal generator must be tuned to the same frequency. The output signal voltage of the generator is reduced to zero and the receiver output noise voltage is read on the ac voltmeter. The output signal voltage of the generator is raised until the reading on the ac voltmeter increases by 10 dB. At this time, the signal generator output voltage must be less than 0.25 microvolt for the receiver to meet its sensitivity specification. An ac voltmeter used in this type of test is calibrated in dB and in voltage; the test is simple and easy to perform. For results that give best accuracy, the ac voltmeter should be of the true rms type.

6–6. *Does an increase in amplification increase sensitivity?*

If a system is lacking in amplification or gain, and noise from the input circuit is not apparent at the output, then amplification would increase the ability of the system to detect small signals. It could be said that sensitivity, in this case, is improved by adding gain. A communications receiver which is so lacking in gain that atmospheric noise cannot be heard at 20 MHz could be improved by adding an amplifier.

If system amplification is sufficient for the noise to be readily apparent at the output, more gain will simply amplify the noise as well as the signals, and no improvement in sensitivity will be obtained. If other measures, such as decreasing bandwidth to the minimum value required or reducing the noise figure of the input circuits, can reduce the noise output, then additional amplification may be of some benefit.

6–7. *How can antennas increase signal-to-noise ratio?*

If the directivity or gain of an antenna can be increased without increasing its source resistance excessively, the signal-to-noise ratio will be improved. Antenna gain may be referred to as noiseless gain, because it can increase signal level without increasing source resistance thermal noise. This characteristic of antenna gain makes large antennas far more desirable than electronic amplifiers, which add circuit noise and decrease signal-to-noise ratio. Signals which are too small to be discerned from the noise when using simple antennas may be detected if antenna gain can be sufficiently increased.

6-8. *What effect does impedance matching have on signal-to-noise ratio?*

Equation 6-1 states that the signal-to-noise ratio of a signal source, such as an antenna, a microphone, or a signal generator, is equal to the squared ratio of the open-circuit signal voltage to the open-circuit thermal noise voltage of the source resistance, or:

$$\text{snr} = \left(\frac{e_s}{e_n}\right)^2$$

$$= \frac{(e_s)^2}{4KTBR_g} \qquad \text{(Eq. 6-6)}$$

Fig. 6-3 shows an equivalent circuit for such a signal source. The thermal noise voltage of source resistance R_g is represented by noise generator e_n, and the source resistor is assumed to be noiseless.

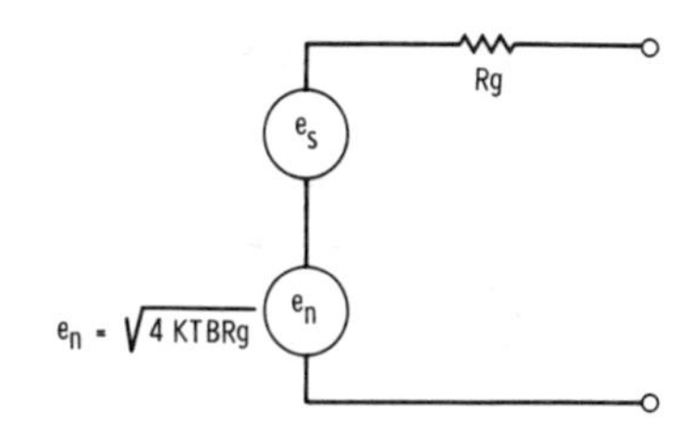

Fig. 6-3. Equivalent circuit of a signal source.

For maximum signal power to be transmitted from the signal source to the load, load resistance R_L, must equal source resistance R_g. The generator and load are then said to be matched. Fig. 6-4 shows a signal source with a matched load, the load resistance being equal to R_g. Also shown is the Thevinin equivalent circuit which indicates the amplitudes of the signal and noise voltages appearing across the load. Notice that the effective resistance across the load terminals is only one-half the value of R_g, because R_g and R_L are effectively in parallel. This reduces the thermal noise voltage by a factor of $1/\sqrt{2}$. The signal voltage appearing across the load terminals is one-half source signal voltage e_s, due to the voltage-divider action of the

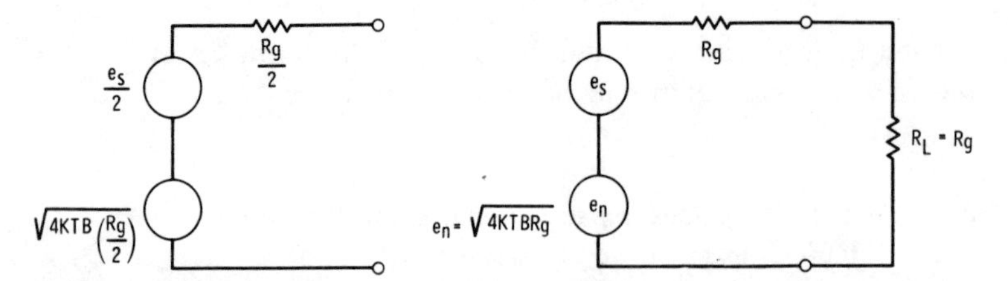

Fig. 6-4. Signal source with matched load (above) and equivalent circuit showing signal and noise voltages across load terminals (below).

two resistors. The value of the signal-to-noise ratio across the load terminals is:

$$\text{snr} = \frac{\left(\frac{e_s}{2}\right)^2}{4KTB\left(\frac{R_g}{2}\right)}$$

$$= \frac{\frac{(e_s)^2}{4}}{2KTBR_g}$$

$$= \frac{(e_s)^2}{8KTBR_g} \qquad \text{(Eq. 6-7)}$$

This is one-half the value of signal-to-noise ratio given by Eq. 6-6 for the open-circuit or unloaded signal source. Thus, the signal-to-noise ratio of a signal source is reduced by one-half, or 3 dB, when a matched load is connected to the terminals of that signal source. This is only true when the noise from the signal source is caused only by the thermal noise of the source resistance.

If the output terminals of a receiver are considered as a signal source, the output resistance of the receiver would have to be considered the source resistance of this signal source. Its thermal noise would be negligible, compared to the noise coming from the input stage of the receiver; therefore, the output signal-to-noise ratio of the receiver would not be affected by connecting a matched load to the receiver output.

6-9. *What is the concept of available noise power?*

The maximum available power, or more simply, the available power, is transmitted to a load when the load resistance is equal to the signal-source resistance. The signal voltage across the load under matched conditions is equal to one-half the open-circuit voltage of the signal source, and the power into the load is:

$$P_s = \frac{\left(\frac{e_s}{2}\right)^2}{R_g}$$

$$= \frac{(e_s)^2}{4R_g} \qquad \text{(Eq. 6-8)}$$

This equation gives the maximum power that the signal source can furnish to a load. If the signal voltage in Eq. 6-8 is replaced by the thermal noise voltage of the source resistance, then the resulting equation expresses the maximum noise power, or the available noise power, that the signal source can supply to a load.

$$P_n = \frac{(e_n)^2}{4R_g}$$

$$= \frac{4KTBR_g}{4R_g}$$

$$= KTB \qquad \text{(Eq. 6-9)}$$

Available noise power, or KTB, has the units of watts, and the terms K, T, and B are the same as used in Eq. 2-1 for thermal noise voltage.

It is interesting to write the equation for signal-to-noise ratio in terms of available signal power and available noise power given in Eq. 6-8 and Eq. 6-9, respectively.

$$snr = \frac{P_s}{P_n}$$

$$= \frac{\frac{(e_s)^2}{4R_g}}{KTB}$$

$$= \frac{(e_s)^2}{4KTBR_g} \qquad \text{(Eq. 6-10)}$$

Equation 6-10 is identical to Eq. 6-6, in which signal-to-noise ratio was determined by using the open-circuit signal and noise voltages of the signal source.

6–10. *Why is KTB, or available noise power, useful?*

Available noise power is a convenient term to use in comparing signal levels to the thermal noise of a source resistance; available noise power is independent of the value of source resistance. The only variables in KTB are temperature and bandwidth. A signal may be expressed as being so many dB more (or less) than KTB, and this expression can be applied to any system, regardless of source resistance. For example, atmospheric or galactic noise is commonly expressed in dB relative to KTB. If atmospheric noise has a value of 30 dB above KTB, then its value in watts is 1000 times greater than KTB, the thermal noise power available from the antenna source resistance. This level of atmospheric noise would induce a voltage in the receiving antenna equal to $\sqrt{1000}$ times the thermal noise voltage of the antenna resistance.

7

Noise Figure

7-1. *What is noise figure?*

Noise figure is a parameter used to describe the noise performance of receivers and amplifiers. It expresses how much the signal-to-noise ratio is decreased by the noise present in the circuits of the receiver or amplifier. If the signal-to-noise power ratio at the input to a receiver is 10 dB, and the signal-to-noise power ratio at the output of this receiver is 6 dB, then the noise figure of the receiver is 4 dB. The signal-to-noise ratio has been reduced 4 dB due to receiver circuit noise. If the input signal-to-noise power ratio is 40 dB, the 4-dB noise figure will result in a signal-to-noise power ratio of 36 dB at the output of the receiver. A hypothetical, ideal receiver would have a noise figure of 0 dB, and the signal-to-noise ratio at the output of the receiver would be the same as at the input. All practical electronic circuits inherently generate noise, however, and practical amplifiers and receivers degrade the signal-to-noise ratio.

Noise figure is the ratio of the input signal-to-noise power ratio to the output signal-to-noise power ratio, and it is commonly expressed in dB. The bandwidth of a system does not have any bearing on its noise figure; therefore, noise figure provides an excellent means of evaluating and comparing the noise performance of different systems.

7-2. *What is noise factor?*

The term noise factor is sometimes used to indicate the value of noise figure before it is converted to dB. It is the numerical ratio of input snr to output snr. To avoid confusion, the term noise figure will be used hereafter to indicate either the numerical value of noise figure or its value expressed in dB.

7-3. *What is the equation for noise figure?*

Noise figure may be expressed mathematically as:

$$NF = \frac{\frac{S_i}{N_i}}{\frac{S_o}{N_o}}$$

$$= \left(\frac{S_i}{N_i}\right)\left(\frac{N_o}{S_o}\right) \qquad \text{(Eq. 7-1)}$$

where,

NF is the numerical value of noise figure,
S_i is input signal power,
N_i is input noise power,
S_o is output signal power,
N_o is output noise power.

It was shown in Part 6 that signal-to-noise power ratios could be replaced with squared voltage ratios if the circuit resistance is the same for both signal and noise voltages. The resistance associated with input signal and noise is the same for both of these quantities, and the resistance associated with output signal and noise is the same for both of those voltages. Thus, Eq. 7-1 may be rewritten as:

$$NF = \left[\frac{(e_s)^2}{(e_n)^2}\right]\left[\frac{(e_{no})^2}{(e_{so})^2}\right]$$

$$= \left[\frac{(e_s)\,(e_{no})}{(e_n)\,(e_{so})}\right]^2 \qquad \text{(Eq. 7-2)}$$

where,

e_s is the open-circuit signal voltage of the signal source in volts,
e_n is the thermal noise voltage of the source resistance, defined in Eq. 2-1, in volts,

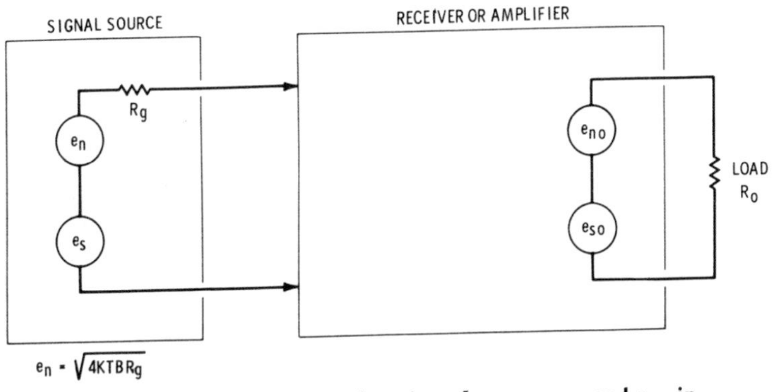

Fig. 7-1. Diagram showing location of e_s, e_n, e_{so}, and e_{no} in equivalent circuit.

e_{so} is the signal voltage across the output load, in volts,
e_{no} is the noise voltage across the output load, in volts.

Fig. 7-1 illustrates where these quantities appear in a system. The output signal and noise voltages are represented by internal voltage generators at the output of the system.

The voltage gain (VG) of the system is equal to the output signal voltage divided by the input signal voltage.

$$VG = \frac{e_{so}}{e_s} \qquad \text{(Eq. 7-3)}$$

This relationship can be substituted in Eq. 7-2 to give:

$$NF = \frac{(e_{no})^2}{(e_n)^2 (VG)^2}$$

$$= \left[\frac{e_{no}}{(e_n)(VG)}\right]^2 \qquad \text{(Eq. 7-4)}$$

Substituting the value of thermal noise voltage from Eq. 2-1 yields:

$$NF = \left[\frac{e_{no}}{(\sqrt{4KTBR_g})(VG)}\right]^2 \qquad \text{(Eq. 7-5)}$$

The numerator in Eq. 7-5 represents the actual output noise power of the system. The denominator represents what the output noise power would be if the receiver or amplifier had no internal noise; it is the thermal noise of the source resistance multiplied by the gain of the system.

7-4. *How is noise figure expressed in dB?*

Two ways of expressing Eq. 7-4 in dB (decibels) are:

$$NF_{dB} = 10 \log_{10} \left[\frac{e_{no}}{(e_n)(VG)} \right]^2 \qquad \text{(Eq. 7-6)}$$

$$NF_{dB} = 20 \log_{10} \left[\frac{e_{no}}{(e_n)(VG)} \right] \qquad \text{(Eq. 7-7)}$$

If the input and output signal and noise voltages are known, noise figure in dB may be calculated as follows:

$$NF = 20 \log_{10} \left(\frac{e_s}{\sqrt{4KTBR_g}} \right) - 20 \log_{10} \left(\frac{e_{so}}{e_{no}} \right) \qquad \text{(Eq. 7-8)}$$

Equation 7-8 is the input signal-to-noise ratio in dB minus the output signal-to-noise ratio in dB.

7-5. *How does impedance matching affect noise figure?*

In Question 6-8 it was shown that signal-to-noise ratio is reduced by one-half (3 dB) when a matched load is connected to the signal source. Since noise figure is defined as the reduction in signal-to-noise ratio, the noise figure of a system must be at least 3 dB if its input resistance matches the source resistance. Due to internal circuit noise in the system, the actual noise figure will be some amount greater than 3 dB.

When a signal source having a source resistance R_g is connected to a load resistance R_L, as in Fig. 7-2, the signal-to-noise

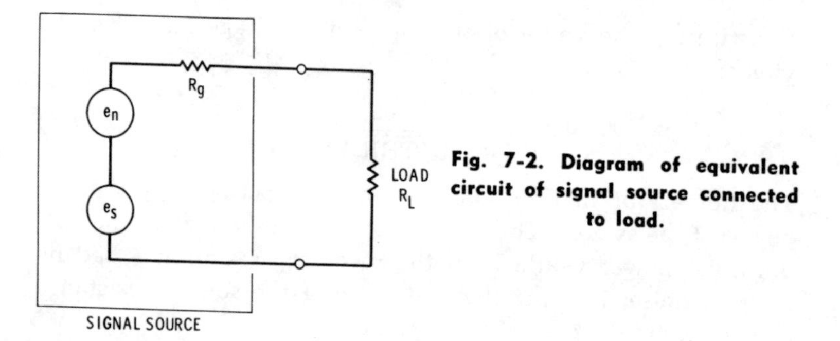

Fig. 7-2. Diagram of equivalent circuit of signal source connected to load.

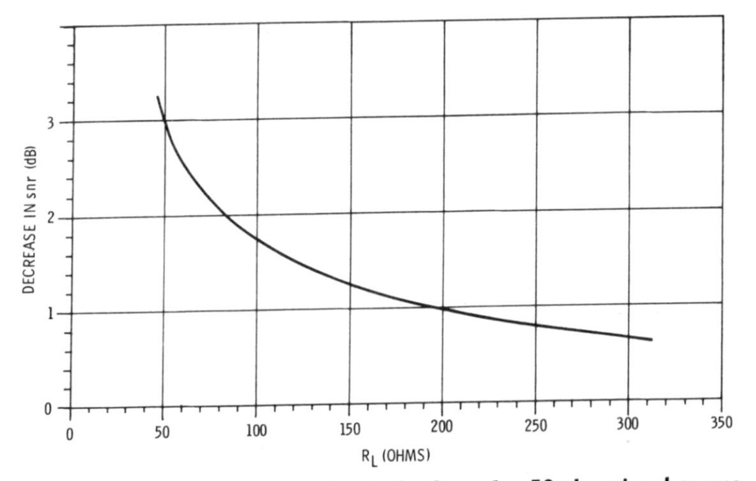

Fig. 7-3. Decrease of input snr due to loading of a 50-ohm signal source.

ratio across the load varies with respect to the load resistance value, as shown in Eq. 7-9.

$$(\text{snr})_{\substack{\text{across}\\\text{load}}} = \left(\frac{R_L}{R_g + R_L}\right)(\text{snr})_{\substack{\text{signal}\\\text{source}}} \qquad \text{(Eq. 7-9)}$$

When R_L is very large compared to R_g, the signal-to-noise ratio across the load will be essentially the same value as that of the signal source. As R_L approaches zero, however, the signal-to-noise ratio across the load will also approach zero. Fig. 7-3 shows how the signal-to-noise ratio varies across a varying load resistance which is connected to a 50-ohm signal source. The vertical axis of the graph indicates how much lower (in dB) the signal-to-noise ratio across the load is than that of the signal source. This curve indicates that noise figure may be improved by making the input resistance of a system larger than the source resistance.

7-6. *How does attenuation affect noise figure?*

The preceding question showed how attenuation due to a voltage divider (formed by R_g and R_L) will reduce signal-to-noise across the load, and thus increase noise figure. Any attenuation which reduces the signal more than it reduces thermal noise will decrease signal-to-noise ratio, which increases noise figure. Losses in the input coupling and tuning circuits of a receiver add to the noise figure of the receiver. Bias resistors may also produce a loss in signal-to-noise ratio and increase noise figure. Best noise figure is obtained by minimizing losses in signal amplitude, particularly in the input circuits of a receiver or amplifier.

7–7. *How is the combined noise figure of several stages of amplification calculated?*

Fig. 7-4 shows a signal source connected to an amplifier. The output of this amplifier is fed to the input of a second amplifier, and a load resistor is connected to the output of the second amplifier. The noise figure of the first amplifier is 3 dB, and the noise figure of the second amplifier is 6 dB. Each amplifier has a power gain of 12 dB (numerical power gain of 16). The overall noise figure of the two amplifiers is given by Eq. 7-10.

$$NF = NF_1 + \frac{NF_2 - 1}{G_1} \qquad \text{(Eq. 7-10)}$$

Subscripts indicate which amplifier the parameters are associated with. NF is noise figure and G is power gain. Numerical values (rather than dB) must be used in Eq. 7-10. The 3-dB noise figure of amplifier 1 is a numerical value of 2, and the 6-dB noise figure of amplifier 2 is a numerical value of 4.

Inserting values from Fig. 7-4 into Eq. 7-10 results in:

$$NF = 2 + \frac{4 - 1}{16}$$

$$= 2 + \frac{3}{16}$$

$$= 2.187$$

When this numerical value of noise figure is converted to dB, the overall noise figure of the two amplifiers is:

$$NF_{dB} = 10 \log_{10} (2.187)$$

$$= 3.4 \text{ dB}$$

Notice that the overall noise figure is not much larger (0.4 dB) than the noise figure of the first amplifier. The equation for the overall noise figure of three amplifiers is:

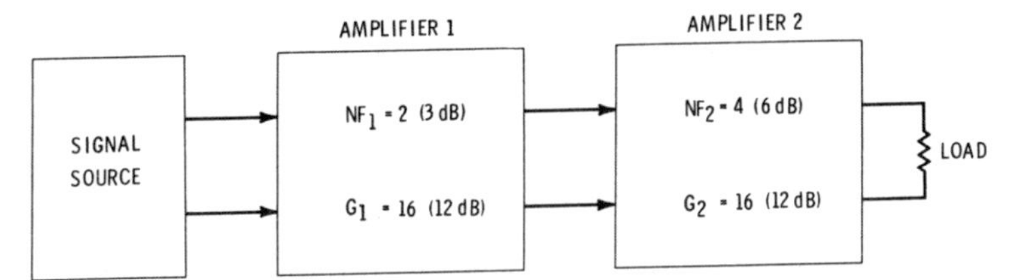

Fig. 7-4. Diagram of signal source connected to load through two amplifiers.

$$NF = NF_1 + \frac{NF_2 - 1}{G_1} + \frac{NF_3 - 1}{(G_1)\,(G_2)} \qquad \text{(Eq. 7-11)}$$

An important fact illustrated by Eq. 7-11 is that the contribution to overall noise figure by each amplifier is decreased by the amount of gain preceding that amplifier. In most cases, the contribution to noise figure by the third amplifier is negligible if the first two amplifiers have reasonable amounts of gain.

7–8. *Why is it important that the first amplifier in a system have high gain and low noise figure?*

If the first amplifier in a system has a high enough gain, the noise of following stages in the system will have very little effect on overall noise figure, even though they may have relatively high noise figures. This is indicated by Eq. 7-10 and Eq. 7-11. Thus, the noise figure of an entire system may be determined by the noise figure of the first stage. Since the noise figure of the first amplifier contributes most to the overall noise figure of the system, it is desirable that it have the lowest possible value of noise figure.

7–9. *How do preamplifiers improve noise figure?*

A preamplifier is an amplifier installed between a signal source and a system. As an example, a preamplifier might be connected between a receiver and its antenna. If the preamplifier has a lower noise figure than the receiver, then the overall noise figure may be lower. For this to be true, however, the gain of the preamplifier must be sufficient to substantially reduce the contribution of the receiver noise figure to overall noise figure (see Question 7-8).

Fig. 7-5 shows a receiving system including an antenna, its transmission line, and a receiver. The signal-to-noise ratio at the antenna is 20 dB. A 6-dB reduction in signal occurs in the transmission line due to losses, and the signal-to-noise ratio at the receiver input is 14 dB. The receiver has a noise figure of 10 dB, which reduces the signal-to-noise ratio to 4 dB at the receiver output.

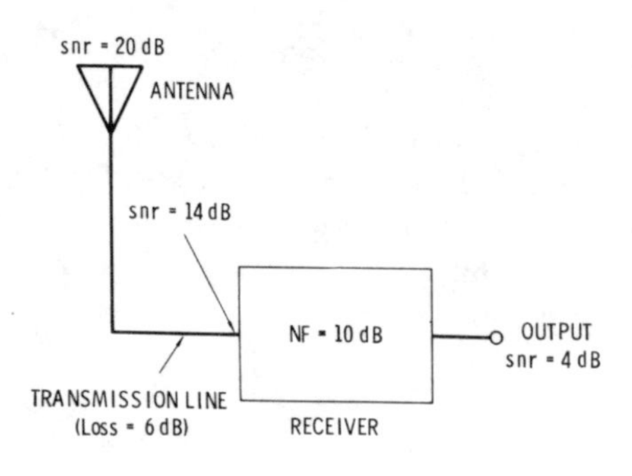

Fig. 7-5. Receiving system.

Fig. 7-6 shows the same system with a low-noise preamplifier installed between the transmission line and the receiver. The preamplifier has a noise figure of 3 dB and a power gain of 12 dB. The signal-to-noise ratio is still 20 dB at the antenna, and it still suffers a 6-dB reduction due to transmission line loss. The overall noise figure of the preamplifier and receiver is calculated (using Eq. 7-10) to be 4.1 dB. The signal-to-noise ratio at the receiver output is, therefore, 9.9 dB. This is nearly a 6-dB

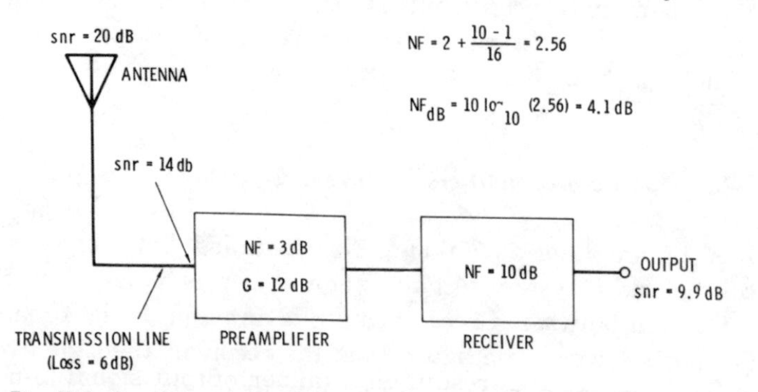

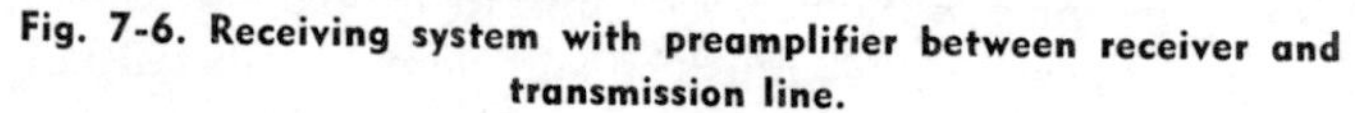

Fig. 7-6. Receiving system with preamplifier between receiver and transmission line.

improvement over the output signal-to-noise ratio obtained when the preamplifier is not used.

If the preamplifier is installed between the antenna and the transmission line, as shown in Fig. 7-7, further improvements in noise figure may be obtained. The 6-dB loss in the transmission line may be added to the 10-dB noise figure of the receiver,

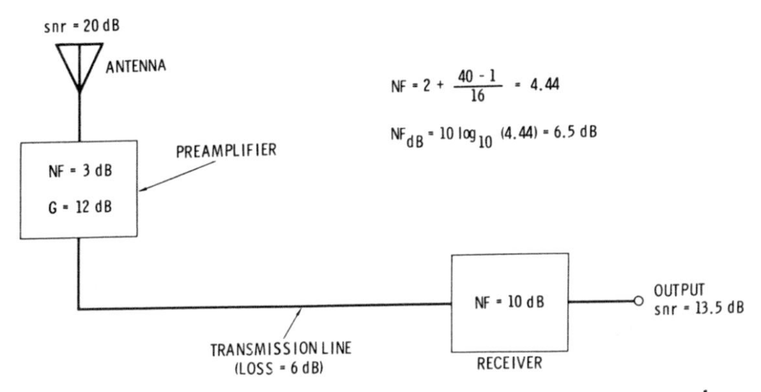

Fig. 7-7. Receiving system with preamplifier between antenna and transmission line.

making a total noise figure for the transmission line and receiver of 16 dB. The numerical value of this noise figure is 40. Using Eq. 7-10, the overall noise figure of preamplifier, transmission line, and receiver is found to be 6.5 dB. Subtracting this from the 20-dB signal-to-noise ratio at the antenna gives the output signal-to-noise ratio of 13.5 dB. Table 7-1 tabulates the output signal-to-noise ratios for the three configurations of Fig. 7-5, Fig. 7-6, and Fig. 7-7.

Table 7-1. Effect of Preamplifier on Output snr

Configuration	Output snr
No Preamplifier	4.0 dB
Preamplifier at Receiver	9.9 dB
Preamplifier at Antenna	13.5 dB

Notice that as the preamplifier is placed nearer the signal source, its gain reduces the effects of loss and noise on the overall noise figure. This results in a higher output signal-to-noise ratio.

7-10. *How can noise figure be measured using a signal generator?*

One way to measure noise figure using a signal generator is indicated by Eq. 7-4. The signal generator is used to measure voltage gain VG of the receiver or amplifier. Equation 7-3 shows that voltage gain is equal to output signal voltage e_{so} divided by open-circuit signal generator voltage e_s. The other two quantities needed to calculate noise figure are output noise voltage e_{no} and the thermal noise voltage e_n, of generator source resistance R_g. Output noise voltage is measured across the output load when source resistance R_g is connected to the input of the receiver or amplifier, but with the signal generator output voltage set at zero. The output load for the receiver or amplifier should be the same when measuring output noise voltage as when measuring voltage gain. Thermal noise voltage e_n of the source resistance may be calculated using Eq. 2-1, but the noise bandwidth of the receiver or amplifier must be known to make this calculation.

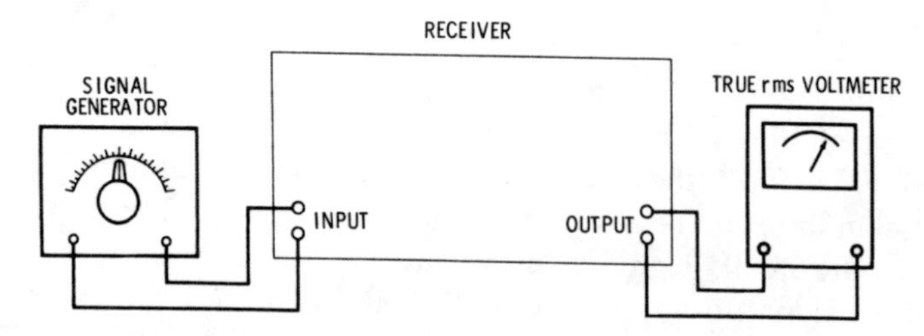

Fig. 7-8. Measuring noise figure with a signal generator.

Another way to measure noise figure using a signal generator is by the arrangement shown in Fig. 7-8. Noise output of the receiver is measured with a true rms voltmeter. This measurement must be made with the signal generator connected to the input or the receiver, but with its output voltage set to zero. The signal generator output is then increased until the reading on the true rms voltmeter increases by a factor of 1.414 (3 dB). This sets the output signal-to-noise ratio at 1.0. Noise figure is then equal to input signal-to-noise ratio (see Eq. 7-1).

$$NF = \frac{(e_s)^2}{4KTBR_g}(1.0)$$

Again, noise bandwidth must be known to calculate the thermal noise voltage of the generator source resistance. It is important to use a true rms voltmeter to ensure that the output signal-to-noise ratio is equal to 1.0 when the reading of the voltmeter has been increased by the factor of 1.414. Average-reading voltmeters introduce significant error when reading noise voltages and voltages which are a combination of noise and signal.

7-11. *How can noise figure be measured with a shot-noise generator?*

A shot-noise generator using a temperature-limited diode was described in Part 3. This type of noise generator is well suited for use in measuring noise figure. Fig. 7-9 shows a typical arrangement of test equipment for making this measurement. Output noise voltage of the system is measured on the voltmeter with the noise generator connected to the system input, but with the noise generator shot-noise output voltage set at zero.

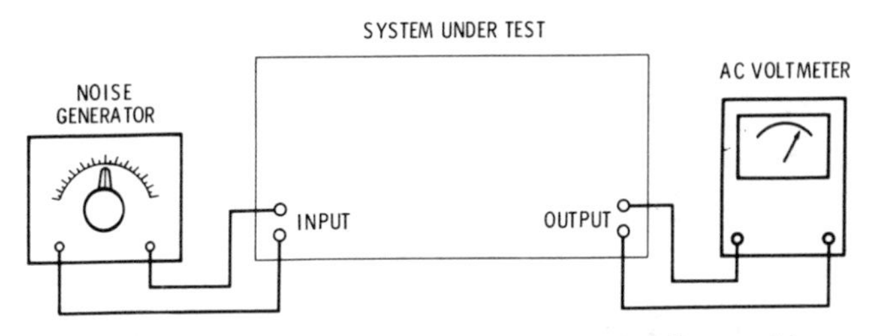

Fig. 7-9. Test arrangement for measuring noise figure with a noise generator.

The output of the noise generator is then raised until the reading on the ac voltmeter increases to 1.414 times the previous reading (3 dB higher). This sets the output signal-to-noise ratio to a value of 1.0 (the shot-noise output of the noise generator is considered to be the signal in this case). Equation 7-1 shows that if the output signal-to-noise ratio is 1.0, noise figure is equal to the input signal-to-noise ratio. The plate current of the noise diode is read from the meter on the noise generator and its value is used to determine noise figure.

The shot-noise voltage (signal) from the noise generator is given by Eq. 3-12:

$$e_{sn} = R_g \sqrt{2eIB}$$

The thermal noise voltage of the generator source resistance is given by Eq. 2-1:

$$e_n = \sqrt{4KTBR_g}$$

These quantities are used to find noise figure in the following manner:

$$NF = \frac{(e_{sn})^2}{(e_n)^2}$$

$$= \frac{2eIB\,(R_g)^2}{4KTBR_g}$$

$$= \frac{eIR_g}{2KT} \qquad \text{(Eq. 7-12)}$$

Notice that noise bandwidth B cancelled out of Eq. 7-12. If a value of 290 K is substituted for temperature in Eq. 7-12, and values are substituted for e and K, Eq. 7-12, reduces to terms of noise diode plate current I and R_g.

$$NF = \frac{(1.6 \times 10^{-19})\,IR_g}{2\,(1.38 \times 10^{-23})\,(290)}$$

$$= 20IR_g \qquad \text{(Eq. 7-13)}$$

In the special case where R_g is 50 ohms,

$$NF = 20\,(I)\,(50)$$

$$= 1000\,(I) \qquad \text{(Eq. 7-14)}$$

If the plate current is expressed in milliamperes, as is usually the case, noise figure is directly equal to current in milliamperes.

$$NF = I_{(mA)} \qquad \text{(Eq. 7-15)}$$

This is the numerical value of noise figure, rather than its value in dB. The value of noise figure may be expressed in dB by use of Eq. 7-16.

$$NF_{dB} = 10 \log_{10}[I_{(mA)}] \qquad \text{(Eq. 7-16)}$$

Table 7-2 shows noise figure in dB versus diode plate current in milliamperes for the case where R_g equals 50 ohms.

The ac voltmeter used to measure output noise voltage need not be a true type for this measurement. An average-reading ac

Table 7-2. Noise Figure in dB Versus Diode Plate Current in mA for R_g = 50 ohms

Noise Figure	Diode Plate Current
1 dB	1.26 mA
2 dB	1.58 mA
3 dB	2.0 mA
4 dB	2.5 mA
6 dB	4.0 mA
8 dB	6.3 mA
10 dB	10.0 mA
12 dB	16.0 mA
13 dB	20.0 mA

voltmeter may be used, because the signal is actually shot noise having the same characteristics as thermal noise. Since an average-reading voltmeter will respond equally to both types of noise, there will be no error involved in its use to set the ratio of two noise voltages at the system output.

7–12. *How is noise figure measured with thermal noise sources?*

Thermal noise generators (resistors) having different temperatures may be used to measure noise figure. Fig. 7-10 shows how such a measurement could be implemented. Two resistors are used as thermal noise voltage sources. Their thermal noise voltages are different because their temperatures are different. The hot resistor may be maintained at a high temperature by means of an oven. The cold resistor may be at room temperature or it may be cooled by some means, such as immersing it in liquid nitrogen. The available noise powers from these resistors

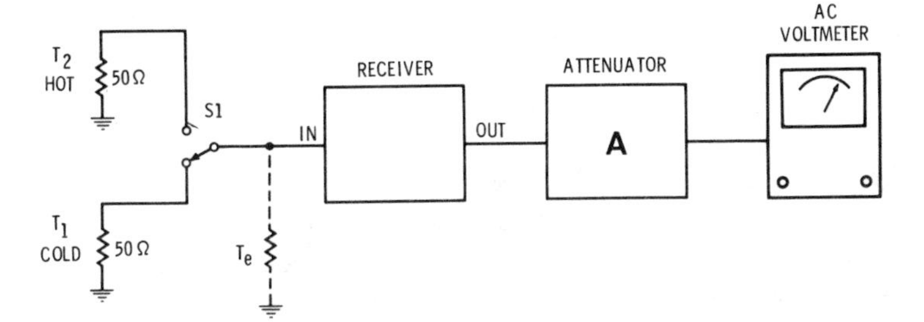

Fig. 7-10. Test arrangement for measuring noise figure with thermal noise sources.

are proportional to their temperatures in degrees Kelvin. That part of the noise power output of the receiver which is due to receiver noise is assumed to be caused by an effective input noise temperature (T_e). This is indicated by dotted lines in Fig. 7-10.

To find the noise figure, switch S1 is set to cold resistor T_1, and the output noise voltage is noted on the ac voltmeter. Switch S1 is then set to hot resistor T_2, which has a higher thermal noise power than the cold resistor. The attenuator is then adjusted until the ac voltmeter reads the same as it did for the cold resistor. This attenuator is calibrated so that it reads attenuation in power ratios greater than unity. For example, if the voltmeter reading doubled when S1 was switched to the hot resistor, the attenuator would have to reduce the receiver output voltage by a factor of 2 to make the voltmeter read the same as it did for the cold resistor. The value of the power ratio is the square of the voltage ratio. In this case, the value of power attenuation A would be equal to 4. From these facts the following equation may be written:

$$A = \frac{T_2 + T_e}{T_1 + T_e} \qquad \text{(Eq. 7-17)}$$

This equation states that the thermal noise power of the hot resistor plus the receiver noise power is A times larger than the thermal noise power of the cold resistor plus the receiver noise power. Solving Eq. 7-17 for T_e results in:

$$T_e = \frac{T_2 - AT_1}{A - 1} \qquad \text{(Eq. 7-18)}$$

The thermal noise power of signal source resistance R_g is proportional to its temperature T_r. Noise figure may be expressed in terms of temperature as:

$$NF = \frac{T_e + T_r}{T_r} \qquad \text{(Eq. 7-19)}$$

The numerator of Eq. 7-19 represents the total noise power of the receiver and the source resistance, while the denominator represents the noise power of only the source resistance. Substituting Eq. 7-18 into Eq. 7-19 gives noise figure as:

$$NF = \frac{T_2 - AT_1}{(A - 1)T_r} + 1 \qquad \text{(Eq. 7-20)}$$

If T_1 and T_r are both at room temperature (290 K), then Eq. 7-20 reduces to:

$$NF = \frac{T_2 - A(290) + 290(A-1)}{(A-1)(290)}$$

$$= \frac{T_2 - 290}{(A-1)(290)}$$

$$= \frac{1}{A-1}\left(\frac{T_2}{290} - 1\right) \qquad \text{(Eq. 7-21)}$$

This gives the numerical value of noise figure. Conversion to dB may be accomplished by use of Eq. 7-22.

$$NF_{dB} = 10 \log_{10}\left[\frac{1}{A-1}\left(\frac{T_2}{290} - 1\right)\right] \qquad \text{(Eq. 7-22)}$$

7-13. *What are some of the advantages and disadvantages of the different methods used to measure noise figure?*

The signal generator method of noise figure measurement is desirable because signal generators are generally available in electronics laboratories. They serve other purposes, such as gain measurement and alignment. This method is most useful when the noise figure to be measured has a rather high value. Signal generators have output power capabilities which are usually very large, compared to thermal noise levels. Perhaps the greatest disadvantage of the signal generator method is that the noise bandwidth of the system measured must be known. A true rms ac voltmeter is required if noise figure is determined by setting the output signal-to-noise ratio to a value of 1.0.

When the shot-noise generator (temperature-limited diode) is used to measure noise figure, it is not necessary to know the noise bandwidth of the system. This saves measurement time and eliminates a source of possible measurement error. Other advantages of this method are: noise figure may be read directly from a meter; the noise generator is a relatively simple piece of equipment; and an average-reading ac voltmeter may be used to indicate output noise voltages. This method of noise figure measurement is not practical, however, for measuring very large values of noise figure. The required value of diode plate current doubles for each additional 3 dB of noise figure value. For example, a shot-noise generator such as shown in Fig. 3-7 ($R_g = 50$ ohms) would require a plate current of 80 milliamperes to measure a noise figure value of 19 dB. The Sylvania 5722 noise diode has a maximum plate current rating of 35 milliamperes.

The thermal noise source method has most of the same advantages as the shot-noise generator method. It is not necessary to know noise bandwidth, and an average-reading ac voltmeter may be used to monitor output noise levels. However, a means must be provided to accurately control resistor temperature at other than room temperature, and a precision attenuator is required.

7–14. *How is the noise figure of a preamplifier measured?*

A preamplifier may not have enough gain to produce a measurable output noise voltage (see Question 5-18). In this case, a postamplifier must be used to amplify the output to a level where noise-figure measurements may be made. A receiver may be used as a postamplifier for rf preamplifiers. To ensure accurate noise figure measurements, the noise figure of the postamplifier should be known. The noise figure of the preamplifier is calculated from the overall noise-figure measurement made on the entire system (preamplifier and postamplifier).

$$NF_{preamp} = NF_{overall} - \frac{NF_{postamp} - 1}{\text{Power Gain of Preamp}} \qquad \text{(Eq. 7-23)}$$

Equation 7-23 is a rearrangement of Eq. 7-10, and numerical values of noise figure and power gain must be used rather than values in dB.

7–15. *What precautions should be observed in measuring noise figure?*

If an electronic system, such as a receiver, has automatic gain control (agc) circuits, they should be disabled so that system gain cannot change during the course of a measurement. If gain measurements are made, the signal level should be adjusted so it is large compared to the system noise, but not so large that clipping or distortion of the waveform occurs. The signal should be in the linear-operating range of the system.

Measurements made using a signal generator require that the frequency of the signal generator be accurately set to the center of the passband. This normally will correspond to the frequency at which maximum gain occurs. True rms ac voltmeters should be used to measure the amplitude of waveforms which are a combination of signal and noise.

Noise figure measurements should be made using signal generators or noise generators having a source resistance equal to the source resistance that the system will see in normal operation. A 50-ohm generator should not be used to make noise figure measurements on a receiver which will be used with a 300-ohm antenna. In this case, a 50- to 300-ohm low-loss impedance transformer should be used between the generator and receiver.

When noise figure measurements are made on receivers, the output noise and/or signal should be measured at the output of the i-f amplifier, ahead of the detector. If the detector is nonlinear (many are), then the signal-to-noise ratio may be altered according to the characteristics of the nonlinearity. Noise figure is a parameter which indicates the noise performance of the receiver input circuits, and distortions of its measurement should not be permitted by nonlinear detectors. If a particular type of detector causes poor noise performance, that problem should be treated separately.

7–16. *What effect do mixers or frequency converters have on noise?*

Mixers used to convert signal frequency to intermediate frequency in superheterodyne receivers tend to have higher noise figures than do well-designed rf amplifiers. This is why low-noise receivers usually have rf amplifiers preceding the first mixer. If the mixer has a linear input-output characteristic, signal-to-noise ratio will not be affected by the mixer (beyond the normal contribution of noise by the mixer). The same may be said of linear product detectors. Nonlinear circuits, however, can change the signal-to-noise ratio in ways not previously discussed.

7-17. *What effect do a-m detectors have on noise?*

Envelope detectors, or amplitude-modulation detectors, can change the signal-to-noise ratio when the signal is small. This is easily demonstrated with an ordinary a-m broadcast receiver. Tune the receiver to a clear, unused frequency and turn up the volume until the noise can be heard. Now tune the receiver to a very weak station and notice the increase in noise as the station is tuned in. Tests have shown that when the noise input to an envelope detector is raised to or above the input signal level, the output signal is decreased or suppressed. This is why it is common practice to make output signal-to-noise ratio measurements on an a-m receiver at the output of the i-f amplier, ahead of the detector.

7-18. *What is spot noise figure?*

A spot noise figure measurement is one in which the noise bandwidth is a small percentage of the center frequency of that bandwidth. The bandwidth should be narrow enough so that the noise figure does not change appreciably within it. Most noise figure measurements made on receivers result in spot noise figures. Spot noise figure must be measured at several frequencies to obtain data on how noise figure changes with frequency.

7-19. *What is average noise figure?*

An average noise figure measurement is one in which the noise bandwidth is so large that the value of noise figure can vary significantly versus frequency within that bandwidth. Thus, the value of noise figure at some particular frequency within the bandwidth may be considerably higher than the average noise figure for the entire bandwidth. Average noise figure is easier to measure at audio frequencies, but it may be less meaningful than a series of spot noise figures measured at different frequencies within the same frequency range.

8

Miscellaneous

8–1. *What is flicker noise?*

Flicker noise is a low-frequency noise which causes noise figure to increase in the lower audio range. It is also called 1/f (one over f) noise, because its influence produces a 3-dB-per-octave rise in noise figure as frequency is decreased. Fig. 8-1 is a typical curve of spot noise figure versus frequency showing how flicker noise can affect the low-frequency performance of an amplifier.

Flicker noise occurs in transistors, FETs, and vacuum tubes. Typical transistor amplifiers may have a 1/f corner frequency (see Fig. 8-1) on the order of 1 kHz or less, depending on the values of collector current and source resistance R_g. The corner frequency tends to shift to lower frequencies at lower values of collector current and higher values of R_g. The same is true of FET amplifiers with respect to R_g, but drain current in FETs may not have nearly as pronounced an effect on noise figure as does collector current in transistors.

8–2. *What is optimum generator resistance?*

Noise figure varies according to the value of generator resistance or source resistance R_g. Optimum generator resistance is that value of R_g at which noise figure has its smallest value. Fig. 8-2 shows a typical curve of transistor noise figure versus R_g. The value of optimum R_g can change for different bias conditions and different frequencies. If transistor manufacturers do not specify the optimum value of R_g at the frequency and bias condition of interest, its value may be found by a series of noise figure measurements using different values of R_g.

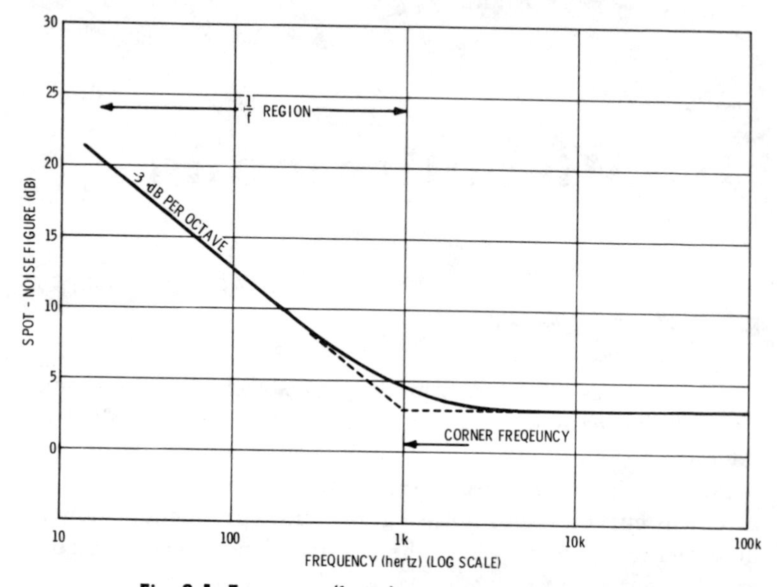

Fig. 8-1. Frequency (hertz) vs spot-noise figure (dB).

Optimum source resistance can be obtained in rf circuits by use of impedance-matching networks to transform the actual source resistance to the desired value. The same may be done in audio circuits by use of audio transformers. It is not a good practice to simply add a resistor in series with a signal source to obtain the optimum value of R_g. Although this procedure may produce the best noise figure for the amplifier, the resulting signal loss and increase in thermal noise may produce a lower output signal-to-noise ratio than was obtained without the added resistor.

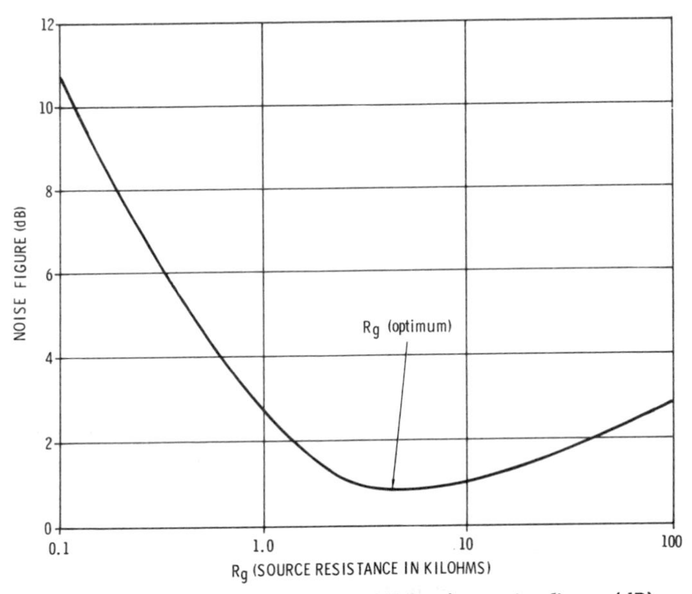

Fig. 8-2. R_g (source resistance, in kilohms) vs noise figure (dB).

8–3. *What are equivalent input noise voltage and noise current* (e_n *and* i_n)?

One method of describing the noise performance of an amplifier is to assume that all of the internal noise of the amplifier can be represented by an imaginary noise voltage generator and an imaginary noise current generator, both located at the input of the amplifier. This is illustrated in Fig. 8-3. These noise generators should not be confused with the thermal and shot-noise generators discussed previously.

If the input terminals of the amplifier are open circuited as shown, noise voltage generator e_n will have no effect, and all of the noise appearing at the amplifier output will be due to noise current generator i_n. When the input terminals are short circuited, however, all of the noise appearing at the amplifier output will be due to noise voltage generator e_n. This is because e_n has zero resistance; all of the noise current from i_n will flow through e_n and none will flow into the amplifier.

This type of noise specification can be applied to many kinds of electronic systems and devices. One or both of these parame-

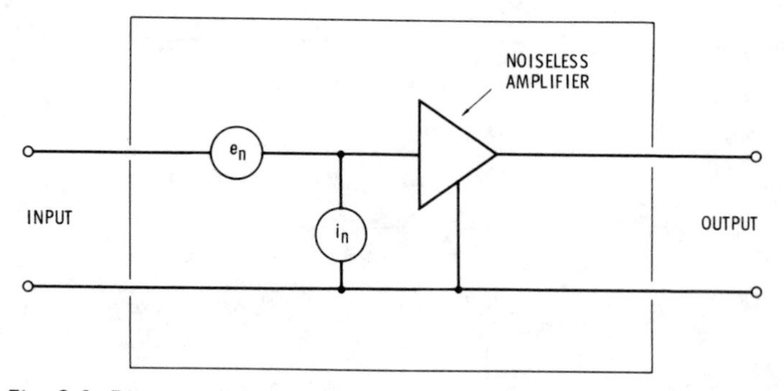

Fig. 8-3. Diagram of imaginary noise voltage generator and imaginary current generator located at the input of an amplifier.

ters may be specified by manufacturers on their transistor and FET data sheets. Commercial test equipment is available to measure their values in transistors and FETs.

8–4. *How may the values of e_n and i_n be used?*

The values of equivalent input noise voltage and current (e_n and i_n) may be used to calculate the optimum value of source resistance R_g for an amplifier or other electronic system. The relationship is:

$$R_{g(optimum)} = \frac{e_n}{i_n} \qquad \text{(Eq. 8-1)}$$

For example, if e_n has a value of 0.4 microvolt, and i_n has a value of 80 picoamperes, the lowest value of noise figure will result when R_g has the value of:

$$R_{g(optimum)} = \frac{0.4 \times 10^{-6}}{80 \times 10^{-12}}$$
$$= 5000 \text{ ohms}$$

8–5. *What is excess noise in resistors?*

Wirewound resistors exhibit only thermal noise, regardless of whether or not current is flowing in them. Their capacitive and inductive characteristics usually limit their use to frequencies below about 50 kHz.

Film and composition resistors, however, show an excess noise (in addition to their thermal noise) which increases with decreasing frequency. This is similar to flicker noise in transistors. Excess resistor noise is usually negligible compared to thermal noise at frequencies above 1.0 MHz. Consideration should be given to the use of wirewound resistors in the first stages of equipment designed for low noise at very low frequencies. Excess resistor noise does not occur in film and composition resistors, unless a current is flowing in the resistor.

8–6. *Do capacitors have noise?*

Most types of capacitors do not display any significant noise level. Tantalum capacitors, used in audio circuits, have occasionally been found to cause large amounts of noise. The cause could not be traced to excessive leakage in these capacitors, but their replacement with other tantalum capacitors eliminated the excessive noise. The use of tantalum capacitors in the input stages of low-noise equipment should be avoided if possible.

8–7. *Do inductors have noise?*

The resistance of the wire used in manufacturing inductors has thermal noise associated with it. In many cases, however, this resistance and its thermal noise are very small compared to other noise sources in the circuit. Unshielded inductors may have a tendency to pick up interference from magnetic fields, so they are vulnerable to man-made noise.

8–8. *What are the noise characteristics of a zener diode?*

Avalanche noise is present in a zener diode when it is biased into its reverse breakdown region. The amplitude of this noise is quite large, compared to thermal and shot-noise sources, and its maximum value appears to occur in the vicinity of the knee of the breakdown curve. See Fig. 8-4.

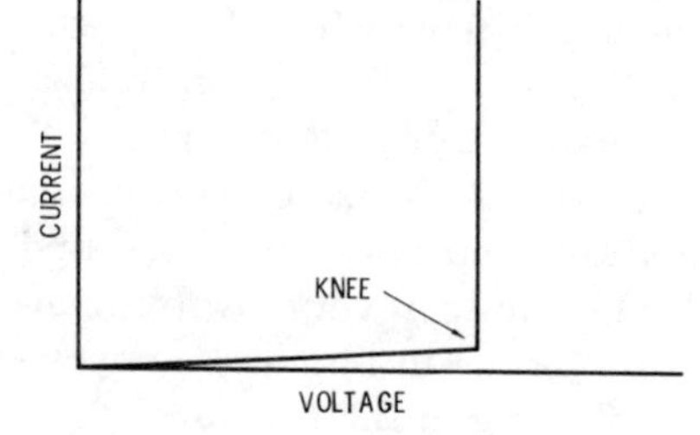

Fig. 8-4. Zener diode breakdown curve.

8–9. *How can zener noise be a problem in low-noise circuits?*

A zener diode may be used to regulate the dc supply voltage to a low-level circuit, as shown in Fig. 8-5. It is easy to see how noise voltage from the zener diode can reach the amplifier input via base-biasing resistor R1. Zener noise will also reach the input of the following amplifier stages via collector load resistor R3.

This problem may be corrected by placing an adequate bypass capacitor across the zener diode, or by using a resistor-capacitor decoupling network between the zener diode and the sensitive circuits.

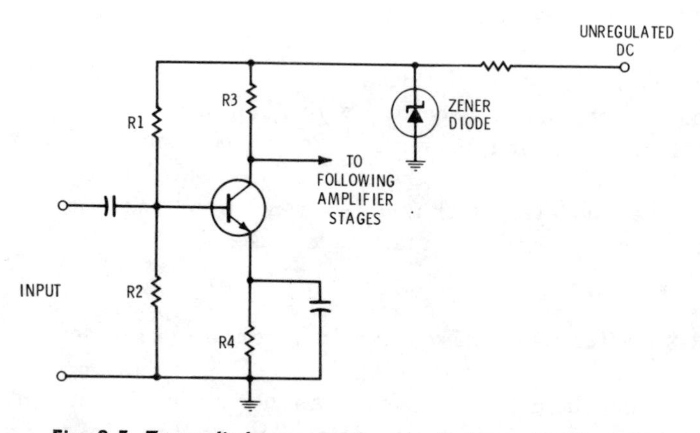

Fig. 8-5. Zener diode can create noise in low-noise circuits.

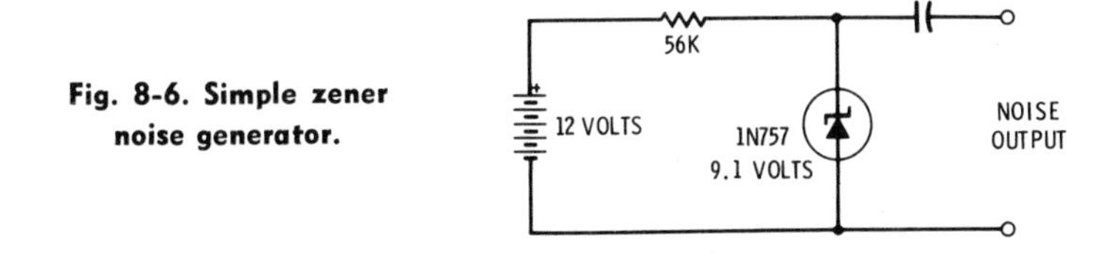

Fig. 8-6. Simple zener noise generator.

8–10. *How can zener noise be used to advantage?*

The relatively high-noise level of zener diodes makes them useful as uncalibrated noise sources in equipment such as radio-frequency bridges and sound synthesizers. Fig. 8-6 shows how a simple zener diode noise source may be constructed. The zener is biased at a current of approximately 52 microamperes.

8–11. *How is the noise rating for vacuum tubes expressed?*

Noise in vacuum tubes is commonly expressed as an equivalent resistance, R_{eq}, which, when connected in the grid circuit of the tube, will produce the same noise current in the plate circuit as normally occurs. The tube is considered noiseless, and all noise is assumed to originate from the thermal noise voltage of R_{eq}. The higher the value of R_{eq}, the more noise will be present in the tube.

In general, noise in vacuum tubes tends to increase with the number of grid elements because of the partition effect. This is a random variation in the way current divides between the positive-biased grids and the plate. R_{eq} for triodes is approximately:

$$R_{eq.} = \frac{2.5}{g_m} \quad \text{(Eq. 8-2)}$$

where,

R_{eq} is in ohms,
g_m is the transconductance of the tube in mhos.

Typical values of R_{eq} for triodes run from several hundred ohms up to a few thousand ohms. Pentodes run higher, being typically 5,000 to 10,000 ohms. When the tube is used as a mixer, or frequency converter, these values of R_{eq} increase substantially. Pentagrid mixers have values of R_{eq} on the order of 200,000 ohms.

Index